T0400025

Robot Learning Human Skills and Intelligent Control Design

Robot Learning Human Skills and Intelligent Control Design

Chenguang Yang

Chao Zeng

Jianwei Zhang

CRC Press
Taylor & Francis Group
Boca Raton London New York

CRC Press is an imprint of the
Taylor & Francis Group, an **informa** business

First edition published 2021
by CRC Press
6000 Broken Sound Parkway NW, Suite 300, Boca Raton, FL 33487-2742

and by CRC Press
2 Park Square, Milton Park, Abingdon, Oxon, OX14 4RN

ISBN: 978-0-367-63436-0 (hbk)
ISBN: 978-0-367-63437-7 (pbk)
ISBN: 978-1-003-11917-3 (ebk)

Typeset in Nimbus Ronam
by KnowledgeWorks Global Ltd.

Contents

Preface

During the last decades, robots are expected to become increasingly intelligent to deal with different types of tasks. More specifically, humans expect that robots could acquire adaptive manipulation skills effectively and efficiently. To this end, several learning algorithms and techniques have been developed and successfully implemented in a number of robotic tasks. Among these methods, learning from demonstrations (LfD) can enable a robot to learn skills more effectively and efficiently from a human tutor. In this way, the robot can be quickly programmed to complete a task. Most of the reported approaches in LfD concentrate on the learning of human movement skills in which only movement/position trajectories are considered and encoded. However, these approaches are hardly applicable for complicated interactions, especially when regarding scenarios where robots physically interact with the environment including humans, and where stiffness/force profiles in addition to position trajectories need to be properly regulated.

In this book, we will introduce our work that can enable robots to efficiently learn both motion control and stiffness/force regulation policies from humans. Both movement trajectories and stiffness profiles extracted from the limb sEMG signals are simultaneously encoded and modeled such that the robot can learn human-like manipulation skills from human demonstrations. We will introduce several machine learning-based algorithms/models that can encode useful information from multimodal demonstration data. We will show that the robot can learn human adaptive skills through skill transfer.

The control strategy is also important for a robot when interacting with its environment. Impedance control which considers force and motion has been widely used in this field. In this book, we will present our work on the transferring of human limb impedance control strategies to the robots such that the adaptive impedance control for the robot can be realized. In addition, some advanced control techniques such as radial basis function neural network control are introduced that can enhance the robot performance of the task execution.

This book will be primarily focused on the transfer of human-like adaptive skills to robots and intelligent control techniques that can enhance the learning performance. The structure of this book is summarized as below.

In the first chapter, we will briefly introduce the method of transmitting human impedance information to the robots. The transfer of human impedance to the robot can enable the robot to perform flexible operations and complete more complex tasks. Then, we will introduce several common methods for modeling human skills which are used to model the teaching trajectories so that the robot can reproduce and generalize the teaching trajectories. The trajectory segmentation technology is also introduced, which will be used for complex multi-step tasks. Next, we will introduce the design of controllers, such as admittance control, variable impedance control, and controller based on neural networks.

Chapter 2 will introduce a variety of robot hardware and software systems to be used in this book. Robot hardware systems include popularly used robots (i.e., Baxter robot, Nao robot, and KUKA LBR iiwa robot), Kinect camera which is an inexpensive RGB-D sensor, MYO Armband, and Leap Motion which can detect and track human hands motion. The introduction of the robot hardware systems includes the composition, important parameters, and functions. Besides, it shows the appearances and application fields of the robot hardware systems. Robot software systems are mainly robot simulation software, including MATLAB Robot Toolbox, CoppeliaSim, and Gazebo.

In Chapter 3, we will focus on how to transfer the mechanism of human stiffness adaptive regulations to a robot. Adaptive control strategies have been shown great potentials in compliant interactions between robots and environments, especially in some physically coupling scenarios. Human-robot stiffness transfer is the key to these "human-like" adaptive control strategies, which enable them to simultaneously adapt force/impedance and motion in the presence of unknown dynamics. In this chapter, we will introduce a number of techniques for human-robot stiffness transfer based on sEMG signals. In these techniques, variable stiffness of demonstrator/tutor is estimated and then transferred to the robot. Several tasks are carried out and show better performances compared with conventional methods.

In Chapter 4, a framework of learning and generalization of variable impedance skills is proposed to help a robot manipulator acquire human-like skills and allows it to adapt to more complex task situations. Several models, such as dynamical movement primitives, are utilized to generalize the learned skills to new given task situations. Significantly, DMP is used to encode both movement trajectories and stiffness profiles under this framework. We combine the DMP with Gaussian Mixture Models to enable the robot to learn from a set of demonstrations of the same task, such that robot can learn from multiple demonstrations and generate a better motion trajectory than that just from one demonstration. Our method is also able to learn and generalize multi-step tasks.

Chapter 5 provides a multimodal teaching-by-demonstration system, which transfers the adaptation of multimodal information from a human tutor to the robot. The hidden semi-Markov model is used to encode the multiple signals in a unified manner and encode the correlations between position and the other three control variables (i.e., velocity, stiffness, and force) separately. The expected control variables are generated by the Gaussian mixture regression based on the estimated parameters of the HSMM model. To reproduce the task, the learned variables are further mapped into an impedance controller in the joint space through inverse kinematics. Finally, the effectiveness of the approach is proved by several experiments.

In Chapter 6, A KF-based sensor fusion is applied to obtain an improved performance, where a Kinect sensor is utilized to capture the motion of the operator's arm with a vector approach. By selecting five out of seven joints on each arm, the vector approach can precisely calculate the angular data of human arm joints. The continuous-time KF method output the designed data with less error; after that, the data will be applied to the joints of a simulated robot, respectively for teleoperation.

A teleoperation system is developed based on visual interaction, and put forward a robot teaching method based on ELM. Kinect is used to control the robot in V-REP by human body motion in the teleoperation system. The robot can reproduce the trajectory which is provided by the RBF network through learning and training.

Chapter 7 will propose a cognitive learning framework for human-robot skills transfer, which simultaneously considers both motion and the contact force during the demonstration. We use the DMP to model the motion and the force to achieve skills generalization. The adaptive admittance model is proposed to simplify the teaching process. To reproduce the motion and the contact force, the hybrid force/motion controller is developed based on the original position controller of the Baxter robot. The NN-based controller is designed to overcome the impact of the unknown payload so that the manipulator is able to track the given motions more accurately.

Author Biography

Prof. Dr. Chenguang Yang received PhD degree from the National University of Singapore and performed postdoctoral research at Imperial College London. He is a recipient of the prestigious IEEE Transactions on Robotics Best Paper Award as lead author and over ten Best Paper Awards from top international conferences. He has been awarded EPSRC Innovation Fellowship and EU FP-7 Marie Curie International Incoming Fellowship. He serves as Associate Editor of a number of leading international journals including IEEE Transactions on Robotics. He is a Co-Chair of the Technical Committee on Collaborative Automation for Flexible Manufacturing (CAFM), IEEE Robotics and Automation Society and Co-Chair of the Technical Committee on Bio-mechatronics and Bio-robotics Systems (B2S), IEEE Systems, Man, and Cybernetics Society.

Dr. Chao Zeng is currently a Research Associate at the Institute of Technical Aspects of Multimodal Systems, Universität Hamburg. He received his Ph.D. degree in robotics and control from South China University of Technology (SCUT) in 2019. During his Ph.D. study, he involved in several scientific research projects and produced several research results that have been published in several journals such as IEEE/ASME Transactions on Mechatronics and IEEE Transactions on Automation Science and Engineering. His research interests include robotic multimodal learning and adaptive compliant manipulation.

Prof. Dr. Jianwei Zhang is the director of TAMS, Department of Informatics, Universität Hamburg, Germany. He received both his Bachelor of Engineering (1986, with distinction) and Master of Engineering (1989) at the Department of Computer Science of Tsinghua University, Beijing, China, his PhD (1994) at the Institute of Real-Time Computer Systems and Robotics, Department of Computer Science, University of Karlsruhe, Germany. His research interests are sensor fusion, intelligent robotics and multimodal machine learning, etc. In these areas, he has published about 400 journal and conference papers, technical reports, and four books. He is the coordinator of the DFG/NSFC Transregional Collaborative Research Centre SFB/TRR169 "Crossmodal Learning" and several EU robotics projects. He has received multiple best paper awards. He is the General Chairs of IEEE MFI 2012, IEEE/RSJ IROS 2015, and the International Symposium of Human-Centered Robotics and Systems 2018. Jianwei Zhang is life-long Academician of Academy of Sciences in Hamburg Germany.

Acknowledgments

Many insightful suggestions and comments from our colleagues, friends, and co-workers have helped us to complete this book. Here, we would like to express our sincere gratitude to them. We would like to express our appreciation to those who have contributed to the collaborative research reported in this monograph. We would like to thank Ying Zhang, Xian Li, Hanzhong Liu, and Chuize Chen who supported for the organization of the materials, and to thank Yuan Guan and Weiyong Si who helped to improve the presentation.

1 Introduction

In this chapter, we will first give a brief review of the state-of-art developments in sEMG-based stiffness transfer from humans to robots. Subsequently, in section 1.2, we will introduce several commonly used approaches that can model human skills and thus enable the skill transfer to robots. In section 1.3, the design of some intelligent controllers that can enhance robotic skill learning will be introduced, such as admittance control, variable impedance control, and neural network-based controllers.

1.1 OVERVIEW OF SEMG-BASED STIFFNESS TRANSFER

It is predicted that humans and robots will be closely working together, sharing workspace, and thus collaborating to fulfil sophisticated tasks in the near future. The experimental results have demonstrated that a human-robot team will be more efficient and flexible than human or robot working alone [1, 2, 3]. Transferring skills from human tutor to robots, especially human impedance adaptive skills [4], is seen as one of the most effective ways to improve the efficiencies of human-robot collaborative systems. Skill transfer is generally defined as the act of learning manipulative skills for the robot according to the demonstration by the human tutor [5, 6].

Research of human motor behavior reveals that human arm can be stabilized mainly using mechanical impedance control during interaction with a dynamic environment [7] which minimizes the interaction force and performance errors. Since motion trajectory transfer could not allow a robot to generate desired stiffness, e.g., the robot manipulator is unable to operate in a compliant or a rigid manner. Therefore it is difficult for the tutor to guarantee compliance, safety, and efficiency of robot manipulation due to variation of tasks and environmental uncertainty. In contrast, impedance control could enable a target relation between force and displacement. It is essential for robots that come in contact with human or the environment to operate in a safe and natural way. The work [7] reported that human central neural system (CNS) is able to adapt endpoint impedance voluntarily when human performs a task. It is thus of great importance to transfer human tutor's adaptive impedance to a robot in dynamic environments to enable human-like compliance and adaptability. A robot could have a better performance by learning human impedance featured skills, such as calligraphic writing [8], welding [9], and switching [10]. Inspired by these research results, biomimetic learning controllers are proposed in Refs. [11, 12, 13] which are able to simultaneously adapt force, impedance, and trajectory in the presence of unknown dynamics. Compared to traditional robotic controllers, they are "human-like" which enabling robots to have some human motor features in an economic perspective and therefore may have great potentials in compliant human-robot interactions, especially in some scenarios with physical contact, e.g., rehabilitation or daily tasks.

There exist different kinds of technologies for transferring human mechanical motion, impedance, or motor control mechanism to robot, e.g., various body sensors and mathematical models. A vision-based model may be a good candidate in transferring human limb movements to the robots [14, 15], but could not transfer force or impedance to the robot such that the transparency among human, robots, and environment may be attenuated. Alternatively, sEMG signals may be ideal bio-signals to incorporate human skills into robots. They reflect human muscles activations that represent human joint motion, force, stiffness etc. [16, 17, 18, 19]. Moreover, sEMG signals are easily accessible and fast adaptive, used in different applications (e.g., rehabilitation, exo-skeleton) coupled with force, sound, or vision sensors [20, 21, 22, 23, 24], etc. Therefore, sEMG signals are widely used for robots to understand human motion intention during implementing tasks.

Generally, sEMG signals can be processed into two divisions: finite class recognition serials and continuous control reference. The former usually refers to pattern recognition, such as hand posture recognition [25, 26] and such data serials are usually used as switch control signals; in contrast the latter refers to extract continuous force, stiffness and even motion serials from sEMG signals which reflect the variations of human limb kinematics and dynamics during limb movement or pose maintenance. Furthermore, the relationship between sEMG and stiffness, force and motion is approximately linear [16], and thus bio-controller design tends to be simple in sEMG-based robot control system. In Ref. [27], sEMG signals are processed to extract incremental stiffness to reduce stiffness estimation error and calibration time. Its application was tested via robot anti-disturbance pose maintenance. In Ref. [28], tele-impedance is implemented via continuous stiffness reference, and in Refs. [29] and [30], finger position from sEMG signals are continuously estimated though with rather relatively large error. In Ref. [31], squaring and low-pass filtering-based signal envelop extraction algorithm, as well as re-sampling method, is employed to extract incremental smooth stiffness from sEMG signals which are then transferred to the robot to mimic human motor behaviors.

As far as humanoid robot manipulator is concerned, it is ideal for transferring human limb dynamic features to the manipulator with elastic actuators because of their geographical similarity inspired by Ref. [13]. There will be many advantages of this human-robot dynamic transfer such as safety, compliant interaction with human, and environment with low contact force, small trajectory errors, and less time consumption [32]. In Ref. [31], sEMG-based writing skill transfer is proposed. Continuous incremental joint stiffness is extracted from sEMG signals during arm implementing tasks and then transferred to the robot arm via a mapping mechanism under robot stable boundaries.

The interactive interface is usually considered as a bridge between the human tutor and a robot for skill transfer. It plays a great role, especially in the scenarios that require impedance regulation skills [33]. Therefore, the interface needs to be carefully considered [34, 35, 36, 37]. Several types of interfaces have been developed to reach this goal. Conventional interfaces such as keyboard, joysticks, or human motion capture device such as Leap Motion, are implemented in most of the simple tasks through programming. But they are not applicable or suitable to complicated interactions involving the sensorimotor feedback. Some emerging techniques, such as virtual reality [38] and augmented reality [39], have been recently introduced into phRI systems as communication interfaces. Generally, such interfaces with sophisticated controllers and sensors are efficient in position control by providing human tutor virtual sensor feedback. However, only position control with virtual feedback is not sufficient for a robot to fulfil complex and flexible tasks. Interactive forces need to be carefully regulated as well in the scenarios where human and robot will inevitably touch each other physically [40, 41].

More recently, the new generation of robot platforms such as the Baxter robot can be taught by using the built-in interfaces provided by the manufacturer in such a way: human tutor leads through the end-effector of the Baxter robot arm by grasping the touchpad on the wrist of the arm. The motion trajectory will be recorded during the lead-through teaching, and then the robot could "playback" the recorded motion. These interfaces may be effective in simple applications where the tasks only depend on position. However, demonstration with tutor's hand while holding the pad could cause unnatural motion and inconvenience since human tutor has to tightly press the pad and operator at the same time. Thus, the tutor's fingers are constrained, and human hand skill transfer to the robot cannot be realized, which may increase human tutor's workload [42]. More importantly, in order to naturally transfer impedance adaptive skills to a robot, it is necessary to physically couple human tutor and the robot, such that human tutor could "feel" the physical interaction with the robot by haptic feedback provided by the robot to the tutor [43]. Therefore, a properly designed coupling interface should be pretty essential for the realization of effective and efficient skill transfer. It has been established that impedance control is of great importance for skill transfer in physical human-robot interaction (phRI) systems. Exploiting sEMG signals extracted from the human upper limb seems to be an effective way to identify the stiffness parameters in the impedance control model. In Ref. [44], a coupling interface is proposed in order to naturally transfer human impedance adaptive skills to a bimanual Baxter robot.

1.2 OVERVIEW OF ROBOT LEARNING MOTION SKILLS FROM HUMANS

In recent decades, robots have been widely applied in both industrial manufacturing and the daily life of individuals. For industrial robots, learning from demonstrations (LfD) [45] is one of the most efficient and straightforward ways to acquire skills that can be directly used in manufacturing. This way is also applicable and even more important for robots operating in the context of daily life, such as the humanoid robot assistants, the robotic prostheses, and the robotic exoskeletons [46]. For humanoid robot assistants, human-like actions that are imitated from the demonstrator make them more friendly to users [47], and the complexity of the global motion planning can be reduced through LfD. For prosthetic manipulators, the control methods usually require continuously updated commands sent from the human; for example, the signals collected from the motor cortex [48] in the brain-computer interface, which heighten the load of the operator. Employing the visual evoked potentials to generate motion commands can reduce this load partly [49]. And to further simplify the control, storing the motions learned from the demonstrations would be an alternative method. However, the flexibility of the robot will be limited if the motions are pre-planned. Hence, it is necessary to develop an effective model for LfD to generalize motions.

In general, there are two approaches of LfD: probability-based approach and dynamic system (DS)-based approach [50, 51]. The probability-based approach seeks to encode the probability distribution of motions in space [52, 53], while the DS-based approach provides more flexible implementation, where a DS is employed to encode the motion profiles. The DS is a powerful tool for motion modeling [54]. Compared with conventional methods, e.g., interpolation techniques, DS offers a flexible solution to model stable and extensible trajectories. Additionally, the motion encoded with the DS is robust to perturbations. An approach based on DS was used to learn stable motion from human demonstrations [55], where the unknown mapping of the DS was approximated using a neural network model named extreme learning machine [56]. The learned model showed adequate stability and generalization. The neural networks (NNs), for instance, extreme learning machine, have been utilized to learn the DS-based model [57, 56, 55]. However, the usage of the NNs complicates the internal structure of the DS and makes the motion learning inefficient. This DS-based method also required considerable demonstration data for training. In contrast, the DMP model, which consists of a nonlinear DS [58], only requires one demonstration to model motion. Here, the DMP models the movement trajectory as a spring-damper system coupled with a nonlinear function. The inherent stability of the spring-damper system enhances the stability and robustness (to perturbations) of the generated motion. The DMP model is represented by a set of differential equations which can encode high dimensional control policies. Thus far, several types of DMP models have been developed and successfully utilized to learn a large number of skills such as Ball-in-a-Cup [59], drumming [59], and serving-water [60]. However, these papers have only utilized the DMP model for encoding movement trajectories.

DMP have been often employed to solve robot learning problems because of their flexibility. In Ref. [61], DMP were modified to model fast movement inherent in hitting motion. Another study used reinforcement learning to combine DMP sequences so that the robot could perform more complex tasks [62]. While both these studies employed multiple DMP to compose a complete action, another study [63] used multiple DMP to model style-adaptive trajectory, where the style of the generated motion could be changed by modulating the weight parameters that were coupled with the goals. In Ref. [64], a method called Compliant Parametric Dynamic Movement Primitives (CPDMP) has extended DMP, enabling it to perform parametric learning on the complex motion. In Ref. [65], a method called DMP Plus has been proposed to achieve lower mean square error and efficient modification at the same time. In Ref. [66], the authors analyzed dynamic motion primitives in the context of a mobile manipulator - a Toyota Human Support Robot (HSR) - and introduced a small extension of dynamic motion primitives that makes it possible to perform whole-body motion with a mobile manipulator. In Ref. [67], a new method has been proposed to endow a robot with the ability of human-robot collaboration and online obstacle avoidance simultaneously. As is mentioned in Ref. [68], an optimal demonstration is difficult to obtain, and multiple demonstrations can encode the ideal trajectory implicitly. Therefore, multiple demonstrations are considered to train DMP model parameter.

Probabilistic approaches have shown good performance in motion encoding [69, 70, 71]. The inherent variability of the demonstrations can be extracted, and thus, more features of the demonstrations can be preserved. In Ref. [72], an LfD framework using a Gaussian mixture model (GMM) and a Bernoulli mixture model was used to extract the features from multiple demonstrations. A new motion was generated through Gaussian mixture regression (GMR). In contrast with the methods mentioned above, GMM combined with GMR can provide additional motion information for robots when learning from multiple demonstrations. In Ref. [54], a learning method named stable estimator of dynamical systems (SEDS) was proposed for motion modeling, where an unknown function was modeled using GMR. DS-GMR is another method that combines the dynamic system with the statistical learning approach [73]. Both methods exploit the robustness and generalization capability of the dynamic system as well as the excellent learning performance of the probabilistic methods.

To take advantage of the performance of the DS and the probabilistic approach, we will integrate DMP and GMM into our proposed system, where the nonlinear function of DMP is modeled with GMM, and its estimate is retrieved through GMR. This modification enables the robot to extract more features of the motions from multiple demonstrations and to generate motions that synthesize these features. The original DMP was learned using the locally weighted regression (LWR) [74], and the locally weighted projection regression [75] was employed to optimize the bandwidth of each kernel of LWR. Despite the added complexity of the learning procedure, these methods enable the DMP to learn from only one demonstration. Reservoir computing (RC) [76] is another method used to approximate the nonlinear function, but its computing efficiency is less than that of GMR.

Inspired by these work and taking into consideration of the fitting performance of the GMM, the Fuzzy GMM (FGMM) [77] is employed to fuse the features of multiple demonstrations into the nonlinear term of the DMP, which has been proposed to improve the learning efficiency of the active curve axis Gaussian mixture model (AcaGMM) [78] and has shown better nonlinearity fitting performance than the conventional GMM. A novel regression algorithm for the FGMM is further developed to retrieve the nonlinear term, according to the geometric significance of the GMR.

However, these papers have only utilized the DMP model for encoding movement trajectories. Recently, stiffness/force adaptation has been investigated based on this model. In Ref. [79], a DMP method has been presented for variable stiffness scheduling, in which the gains are represented using a function approximator [80], a DMP model is developed for the execution of bimanual or tightly coupled cooperative tasks by coupling force/torque feedback into the transform system. In Ref. [41], the authors proposed a method to simultaneously transfer positional and force requirements for in-contact tasks. In Ref. [81], stiffness adaptation has been intro-duced to assist human tutor to increase the intuitiveness of interaction during teaching. In Ref. [82], the authors proposed a method that enables stiffness adaptation in path operational, considering the variance of motion across multiple executions and the current speed. These papers have verified the significance of stiffness/force adaptation for robot interactions with the environment or humans. However, it has not been investigated to model and learn the stiffness profiles in "muscle" space. This paper proposes a framework that enables robots to efficiently learn both motor and stiffness control policies from humans. To that end, human limb muscle activities are monitored for variable stiffness estimation. The developed framework can simultaneously encode both trajectories and stiffness profiles in a unified manner, which allows both trajectory generalization and stiffness schedule.

For complex multi-step tasks, it is not easy to well generalize the learned policies well and adapt them to new tasks. One promising way to address this problem is to segment the learned skill into a set of sub-skills. Each sub-skill can be then individually modeled and generalized in order to fulfil the requirements of a specific task, and it can also be easily reused in other similar tasks. However, skill segmentation by hand is faced with a large number of problems such as the lack of flexibility, low efficiency, and heavy workload. Most importantly, to divide a skill into a set of sub-skills is often difficult since an effective segmentation technique often requires the knowledge of the robot's kinematic properties [83]. Recently, the Beta Process Autoregressive HMM (BP-AR-HMM) has been proposed in Ref. [84], which can automatically divide multi-model human behaviors into a sequence of features. The work [83, 85] has introduced this algorithm in a LfD framework to sparse the demonstrated trajectories into several sequences. This model has a number of merits, such as (i) no *prior* knowledge is in advance specifically required for the selection of the number of the features. This largely facilitates the segmentation process and makes it possible to potentially build a repository with an infinite number of skills, and (ii) it is able to discover the same features from multiple demonstrations. This is of great importance because it is often necessary to teach a robot several times for a specific task. The work [86] takes one step forward from [85]: the estimated human stiffness profiles from demonstrations are segmented along with the movement trajectories, and thus both the stiffness profiles and movement trajectories can be considered in the latter modeling process.

1.3 OVERVIEW OF ROBOT INTELLIGENT CONTROL DESIGN

In the teaching phase of LfD, admittance control has been widely used in human-robot interaction, as it can generate robot motions based on the human force [87, 88, 89]. Thus, in this book, we use the admittance control to achieve the human-guided teaching. The admittance control exploits the end-effector position controller to track the output of an admittance model. Most studies on admittance control have not considered the human factor, which is an important part of this control loop. The interaction force between robot and human can be used to recognize the human intent and to further improve the interaction safety and user experience [90]. In Ref.

[3], the human force was employed to compute the desired movement trajectory of the human, which is used in the performance index of the admittance model. In Ref. [91], the unknown human dynamics was considered in the control loop. In Ref. [92], an adaptive admittance controller is developed to take into account the unknown human dynamics so that the human tutor can smoothly move the robot around in the teaching phase. The human transfer function and the admittance model were formulated as a Wiener filter, and a task model was used to estimate the human intent. The Kalman filter estimate law was used to tune the parameters of the admittance model. However, the task model in this work is assumed as a certain linear system, which is unreasonable because the estimated human motions should be different for each individual due to different motion habits. Thus, a task model that involves human characteristics needs to be developed. GMR is an effective algorithm to encode human characteristics based on human demonstrations [69, 72]. In Ref. [73], the human motion was analyzed using a Gaussian mixture model, and a new motion that involves the human motion distribution was generated using GMR. This algorithm shows a great feasibility to develop a task model that involves human characteristics.

Some advanced adaptive control strategies have recently been proved effective for the improvement of the control performance [93, 94, 95]; thus, it would be necessary to develop adaptive control techniques in LfD systems. One possible way is to capture the features of human limb stiffness adaptation during a specific task, and then transfer these features to a robot. For instance, a robot can perform tasks in a human-like manner by learning adaptive impedance control [12, 96, 97, 98]. Particularly, transferring human arm signals to a robot can achieve variable impedance control for the robot arm, which has demonstrated a better performance than without human-to-robot stiffness transfer [99, 100]. According to Refs. [101, 102, 103, 104], force profiles in addition to positional profiles should also be regulated, especially for in-contact tasks. sEMG-based variable impedance transfer is able to enable the robot to adapt the force profiles to different task situations with a natural regulation process [105, 100].

In the learning phase of LfD, the robotic motion caused by the human guidance will be modelled according to the modeling method introduced in Section 1.2. However, only considering the motion modeling is not sufficient for a stable LfD framework because of the dynamic and unstructured environments, which will result in many disturbances and the variation of the robot's dynamics. The generated motion is finally used in reproduction, the accuracy of which depends on the performance of the trajectory tracking controller. The controller design methods can be classified into model-based methods and model-free methods. The model-based method has better tracking accuracy because the robot dynamics is considered. However, an accurate dynamic model of a manipulator cannot be obtained in advance due to some uncertainties, e.g., unknown payload. In addition, if the manipulator is controlled using a model-based control method [106], the situation mentioned above will affect the control performance and even make the system unstable. Considering the uncertainties of the robot dynamics, the approximation-based controllers have been designed to overcome such uncertainties. They utilize function approximation tools to learn the nonlinear characteristics of the robot dynamics, various approximation tools such as NNs [107, 97, 108] and fuzzy logic systems [109] have been integrated into the control design to approximate the uncertainties.

Recently, NN has served as a promising computational tool in various fields; for example, the primal-dual neural network has been employed to solve a complicated quadratic programming problem [110]. In Ref. [111], the backpropagation NN (BPNN) was employed to approximate the unknown nonlinear function in the model of the vibration suppression device, while in Ref. [112], the radial basis function NN (RBFNN) was utilized to approximate the unknown nonlinearity of the telerobot system. The work [92] use the RBFNN to approximate the robot dynamics so that the robot can complete the reproduced motion accurately without the knowledge of the robot manipulator dynamics. Compared to BPNN, the learning procedure of RBFNN is based on local approximation; thus, RBFNN can avoid getting stuck in the local optimum and has a faster convergence rate. Besides, the number of hidden layer units of RBFNN can be adaptively adjusted during the training phase, making NN more flexible and adaptive. Therefore, RBFNN is more appropriate for the design of real-time control. For the dynamics controllers that employ the NNs, the learning efficiency is an important aspect that should be considered, because of the trade-off between the approximation accuracy and the efficiency of the NNs. The cerebellar model articulation controller (CMAC) is a type of NNs that have been adopted widely in dynamics control design [113, 114, 115, 116]. The structure of the CMAC is inspired by the information processing model of the cerebellum [115]. This NN is not fully connected to associative memory; thus, local weights are updated during each learning cycle to provide faster learning compared to fully connected NNs, without function approximation loss [114]. The work [117] has also developed a CMAC-NN-based controller to guarantee that the generated motions can be performed accurately and steadily under the output constraint. This constraint exists commonly in real-world robotic systems such as nonholonomic mobile robots (NMRs) [118], and its effect can be compensated with the help of a Barrier Lyapunov Function (BLF). The CMAC is employed to approximate the unknown dynamics of the robot. Therefore, the CMAC-NN could be integrated into the controller to cognize the dynamic environment and to compensate for the unknown dynamics, whereby the robot was able to track the trajectories generated from the motion model more accurately. The work [119] proposed a complete robot learning framework. The DMP, combined with the FGMM, is used to model the demonstrations in the Cartesian space. Then the generated motions are transformed into the trajectories in joint space using the inverse kinematics, and a CMAC-NN-based controller is developed to track the trajectories.

REFERENCES

1. Jörg Krüger, Terje K Lien, and Alexander Verl. Cooperation of human and machines in assembly lines. *CIRP Annals-Manufacturing Technology*, 58(2):628–646, 2009.
2. Hao Ding, Malte Schipper, and Björn Matthias. Optimized task distribution for industrial assembly in mixed human-robot environments-case study on IO module assem-

bly. In *2014 IEEE International Conference on Automation Science and Engineering (CASE)*, pages 19–24. IEEE, 2014.

3. Isura Ranatunga, Sven Cremer, Dan O Popa, and Frank L Lewis. Intent aware adaptive admittance control for physical human-robot interaction. In *2015 IEEE International Conference on Robotics and Automation (ICRA)*, pages 5635–5640. IEEE, 2015.

4. Arash Ajoudani. *Transferring Human Impedance Regulation Skills to Robots*. Springer, 2016.

5. Karinne Ramirez-Amaro, Michael Beetz, and Gordon Cheng. Transferring skills to humanoid robots by extracting semantic representations from observations of human activities. *Artificial Intelligence*, 2015.

6. Bidan Huang, Miao Li, Ravin Luis De Souza, Joanna J Bryson, and Aude Billard. A modular approach to learning manipulation strategies from human demonstration. *Autonomous Robots*, 40(5):903–927, 2016.

7. Etienne Burdet, Rieko Osu, David W Franklin, Theodore E Milner, and Mitsuo Kawato. The central nervous system stabilizes unstable dynamics by learning optimal impedance. *Nature*, 414(6862):446–449, 2001.

8. Toru Tsumugiwa, Ryuichi Yokogawa, and Kei Hara. Variable impedance control based on estimation of human arm stiffness for human-robot cooperative calligraphic task. In *Robotics and Automation, 2002. Proceedings. ICRA'02. IEEE International Conference on*, volume 1, pages 644–650. IEEE, 2002.

9. Mustafa Suphi Erden and Aude Billard. Hand impedance measurements during interactive manual welding with a robot. *Robotics, IEEE Transactions on*, 31(1):168–179, 2015.

10. Daniel S Walker, J Kenneth Salisbury, and Günter Niemeyer. Demonstrating the benefits of variable impedance to telerobotic task execution. In *Robotics and Automation (ICRA), 2011 IEEE International Conference on*, pages 1348–1353, May 2011.

11. Etienne Burdet, Gowrishankar Ganesh, Chenguang Yang, and Alin Albu-Sch?ffer. Interaction force, impedance and trajectory adaptation: By humans, for robots. In Oussama Khatib, Vijay Kumar, and Gaurav Sukhatme, editors, *Experimental Robotics*, volume 79 of *Springer Tracts in Advanced Robotics*, pages 331–345. Springer Berlin Heidelberg, 2014.

12. Chenguang Yang, Gowrishankar Ganesh, Sami Haddadin, Sven Parusel, Alin Albu-Schaeffer, and Etienne Burdet. Human-like adaptation of force and impedance in stable and unstable interactions. *Robotics, IEEE Transactions on*, 27(5):918–930, Oct 2011.

13. Chenguang Yang, Zhijun Li, and E. Burdet. Human like learning algorithm for simultaneous force control and haptic identification. In *Intelligent Robots and Systems (IROS), 2013 IEEE/RSJ International Conference on*, pages 710–715, Nov 2013.

14. Weibo Song, Xianjiu Guo, Fengjiao Jiang, Song Yang, Guoxing Jiang, and Yunfeng Shi. Teleoperation humanoid robot control system based on Kinect sensor. In *Intelligent Human-Machine Systems and Cybernetics (IHMSC), 2012 4th International Conference on*, volume 2, pages 264–267, Aug 2012.

15. Christopher Stanton, Anton Bogdanovych, and Edward Ratanasena. Teleoperation of a humanoid robot using full-body motion capture, example movements, and machine learning.

16. Rieko Osu, David W. Franklin, Hiroko Kato, Hiroaki Gomi, Kazuhisa Domen, Toshinori Yoshioka, and Mitsuo Kawato. Short- and long-term changes in joint co-contraction associated with motor learning as revealed from surface EMG. *Journal of Neurophysiology*, 88(2):991–1004, 2002.

17. GC Ray and SK Guha. Relationship between the surface E.M.G. and muscular force. *Medical and Biological Engineering and Computing*, 21(5):579–586, 1983.
18. Jörn Vogel, Claudio Castellini, and Patrick van der Smagt. EMG-based teleoperation and manipulation with the DLR LWR-III. In *Intelligent Robots and Systems (IROS), 2011 IEEE/RSJ International Conference on*, pages 672–678, Sept 2011.
19. Kyung-Jin You, Ki-Won Rhee, and Hyun-Chool Shin. Finger motion decoding using EMG signals corresponding various arm postures. *Experimental neurobiology*, 19(1):54–61, 2010.
20. Ji Won Yoo, Dong Ryul Lee, Yon Ju Sim, Joshua H You, and Cheol J Kim. Effects of innovative virtual reality game and EMG biofeedback on neuromotor control in cerebral palsy. *Bio-medical materials and engineering*, 24(6):3613–3618, 2014.
21. G Fabian Volk, M Finkensieper, and O Guntinas-Lichius. [EMG biofeedback training at home for patient with chronic facial palsy and defective healing]. *Laryngo-Rhino-Otologie*, 93(1):15–24, 2014.
22. Arash Ajoudani, Nikolaos G Tsagarakis, and Antonio Bicchi. Tele-impedance: Preliminary results on measuring and replicating human arm impedance in tele operated robots. In *Robotics and Biomimetics (ROBIO), 2011 IEEE International Conference on*, pages 216–222, Dec 2011.
23. Ning Wang, Chenguang Yang, Michael R Lyu, and Zhijun Li. An EMG enhanced impedance and force control framework for telerobot operation in space. In *Aerospace Conference, 2014 IEEE*, pages 1–10. IEEE, 2014.
24. Atau Tanaka and R Benjamin Knapp. Multimodal interaction in music using the electromyogram and relative position sensing. In *Proceedings of the 2002 conference on New interfaces for musical expression*, pages 1–6. National University of Singapore, 2002.
25. Jun-Uk Chu, Inhyuk Moon, and Mu-Seong Mun. A real-time EMG pattern recognition system based on linear-nonlinear feature projection for a multifunction myoelectric hand. *Biomedical Engineering, IEEE Transactions on*, 53(11):2232–2239, 2006.
26. Mahdi Khezri and Mehran Jahed. Real-time intelligent pattern recognition algorithm for surface EMG signals. *Biomedical engineering online*, 6(1):45, 2007.
27. Peidong Liang, Chenguang Yang, Ning Wang, Zhijun Li, Ruifeng Li, and Etienne Burdet. Implementation and test of human-operated and human-like adaptive impedance controls on Baxter robot. In *Advances in Autonomous Robotics Systems*, pages 109–119. Springer, 2014.
28. Arash Ajoudani, Nikolaos G Tsagarakis, and Antonio Bicchi. Tele-impedance: Towards transferring human impedance regulation skills to robots. In *Robotics and Automation (ICRA), 2012 IEEE International Conference on*, pages 382–388, May 2012.
29. Ryan J Smith, Francesco Tenore, David Huberdeau, RE Cummings, and Nitish V Thakor. Continuous decoding of finger position from surface EMG signals for the control of powered prostheses. In *Engineering in Medicine and Biology Society, 2008. EMBS 2008. 30th Annual International Conference of the IEEE*, pages 197–200. IEEE, 2008.
30. Jimson G Ngeo, Tomoya Tamei, and Tomohiro Shibata. Continuous and simultaneous estimation of finger kinematics using inputs from an EMG-to-muscle activation model. *J Neuroeng Rehabil*, 11(122):0003–11, 2014.

31. Peidong Liang, Chenguang Yang, Zhijun Li, and Ruifeng Li. Writing skills transfer from human to robot using stiffness extracted from sEMG. pages 19–24, 2015.

32. Arash Ajoudani, Nikolaos G Tsagarakis, and Antonio Bicchi. Tele-impedance: Teleoperation with impedance regulation using a body-machine interface. *The International Journal of Robotics Research*, page 0278364912464668, 2012.

33. Ellen Klingbeil, Samir Menon, Keegan Go, and Oussama Khatib. Using haptics to probe human contact control strategies for six degree-of-freedom tasks. In *Haptics Symposium (HAPTICS), 2014 IEEE*, pages 93–95, Feb 2014.

34. Sami Haddadin, Alin Albu-Schäffer, and Gerd Hirzinger. Safe physical human-robot interaction: measurements, analysis & new insights. In *International symposium on robotics research (ISRR2007), Hiroshima, Japan*, pages 439–450, 2007.

35. Michael A Goodrich and Alan C Schultz. Human-robot interaction: a survey. *Foundations and trends in human-computer interaction*, 1(3):203–275, 2007.

36. Vincent Duchaine and Clément Gosselin. Safe, stable and intuitive control for physical human-robot interaction. In *Robotics and Automation, 2009. ICRA '09. IEEE International Conference on*, pages 3383–3388, May 2009.

37. Emanuele Magrini, Fabrizio Flacco, and Alessandro De Luca. Control of generalized contact motion and force in physical human-robot interaction. In *Robotics and Automation (ICRA), 2015 IEEE International Conference on*, pages 2298–2304, May 2015.

38. Giuseppe Di Gironimo, Giovanna Matrone, Andrea Tarallo, Michele Trotta, and Antonio Lanzotti. A virtual reality approach for usability assessment: case study on a wheelchair-mounted robot manipulator. *Engineering with Computers*, 29(3):359–373, 2013.

39. Rong Wen, Wei-Liang Tay, Binh P Nguyen, Chin-Boon Chng, and Chee-Kong Chui. Hand gesture guided robot-assisted surgery based on a direct augmented reality interface. *Computer methods and programs in biomedicine*, 116(2):68–80, 2014.

40. Petar Kormushev, Sylvain Calinon, and Darwin G Caldwell. Imitation learning of positional and force skills demonstrated via kinesthetic teaching and haptic input. *Advanced Robotics*, 25(5):581–603, 2011.

41. Franz Steinmetz, Alberto Montebelli, and Ville Kyrki. Simultaneous kinesthetic teaching of positional and force requirements for sequential in-contact tasks. In *Humanoid Robots (Humanoids), 2015 IEEE-RAS 15th International Conference on*, pages 202–209. IEEE, 2015.

42. TM Lam, M Mulder, and MM Van Paassen. *Stiffness-force feedback in UAV teleoperation*. In-Tech Education and Publishing, 2009.

43. Etienne Burdet, Gowrishankar Ganesh, Chenguang Yang, and Alin Albu-Schäffer. Interaction force, impedance and trajectory adaptation: by humans, for robots. In *Experimental Robotics*, pages 331–345. Springer Berlin Heidelberg, 2014.

44. Chenguang Yang, Chao Zeng, Peidong Liang, Zhijun Li, Ruifeng Li, and Chunyi Su. Interface design of a physical human robot interaction system for human impedance adaptive skill transfer. *IEEE Transactions on Automation Science and Engineering*, 15(1):329–340, 2018.

45. Stefan Schaal. Learning from demonstration. In *Proc. Adv. Neural Inf. Process. Syst.*, pages 1040–1046, 1997.

46. Zhijun Li, Bo Huang, Arash Ajoudani, Chenguang Yang, Chun-Yi Su, and Antonio Bicchi. Asymmetric bimanual control of dual-arm exoskeletons for human-cooperative manipulations. *IEEE Trans. Robot.*, 34(1):264–271, 2018.

47. Kerstin Dautenhahn, Sarah Woods, Christina Kaouri, Michael L Walters, Kheng Lee Koay, and Iain Werry. What is a robot companion-friend, assistant or butler? In *Proc. IEEE/RSJ Int. Conf. Intell. Robots Syst.*, pages 1192–1197, 2005.

48. Meel Velliste, Sagi Perel, M Chance Spalding, Andrew S Whitford, and Andrew B Schwartz. Cortical control of a prosthetic arm for self-feeding. *Nature*, 453(7198):1098–1101, 2008.

49. Suna Zhao, Zhijun Li, Rongxin Cui, Yu Kang, Fuchun Sun, and Rong Song. Brain–machine interfacing-based teleoperation of multiple coordinated mobile robots. *IEEE Trans. Ind. Electron.*, 64(6):5161–5170, 2017.

50. Jianbing Hu, Zining Yang, Zhiyang Wang, Xinyu Wu, and Yongsheng Ou. Neural learning of stable dynamical systems based on extreme learning machine. In *2015 IEEE International Conference on Information and Automation*, pages 306–311, 2015.

51. Xiaochuan Yin and Qijun Chen. Learning nonlinear dynamical system for movement primitives. In *2014 IEEE International Conference on Systems, Man, and Cybernetics (SMC)*, pages 3761–3766, 2014.

52. A.J. Ijspeert, Jun Nakanishi, Heiko Hoffmann, Peter Pastor, and Stefan Schaal. Dynamical movement primitives: Learning attractor models for motor behaviors. *Neural computation*, 25, 11 2012.

53. Sylvain Calinon, Zhibin Li, Tohid Alziadeh, Nikos Tsagarakis, and Darwin Caldwell. Statistical dynamical systems for skills acquisition in humanoids. pages 323–329, 11 2012.

54. S Mohammad Khansari-Zadeh and Aude Billard. Learning stable nonlinear dynamical systems with Gaussian mixture models. *IEEE Trans. Robot.*, 27(5):943–957, 2011.

55. Jianghua Duan, Yongsheng Ou, Jianbing Hu, Zhiyang Wang, Shaokun Jin, and Chao Xu. Fast and stable learning of dynamical systems based on extreme learning machine. *IEEE Trans. Syst., Man, Cybern. A, Syst.*, to be published, doi: 10.1109/TSMC.2017.2705279.

56. Chenguang Yang, Kunxia Huang, Hong Cheng, Yanan Li, and Chun-Yi Su. Haptic identification by ELM-controlled uncertain manipulator. *IEEE Trans. Syst., Man, Cybern. A, Syst.*, 47(8):2398–2409, 2017.

57. Jianbing Hu, Zining Yang, Zhiyang Wang, Xinyu Wu, and Yongsheng Ou. Neural learning of stable dynamical systems based on extreme learning machine. In *Information and Automation, 2015 IEEE International Conference on*, pages 306–311. IEEE, 2015.

58. Auke Jan Ijspeert, Jun Nakanishi, Heiko Hoffmann, Peter Pastor, and Stefan Schaal. Dynamical movement primitives: learning attractor models for motor behaviors. *Neural Comput.*, 25(2):328–373, 2013.

59. Aleš Ude, Andrej Gams, Tamim Asfour, and Jun Morimoto. Task-specific generalization of discrete and periodic dynamic movement primitives. *IEEE Transactions on Robotics*, 26(5):800–815, 2010.

60. Heiko Hoffmann, Peter Pastor, Dae-Hyung Park, and Stefan Schaal. Biologically-inspired dynamical systems for movement generation: Automatic real-time goal adaptation and obstacle avoidance. In *2009 IEEE International Conference on Robotics and Automation*, pages 2587–2592, May 2009.

61. Katharina Mülling, Jens Kober, Oliver Kroemer, and Jan Peters. Learning to select and generalize striking movements in robot table tennis. *Int. J. Robot. Res.*, 32(3):263–279, 2013.

62. Freek Stulp, Evangelos A Theodorou, and Stefan Schaal. Reinforcement learning with sequences of motion primitives for robust manipulation. *IEEE Trans. Robot.*, 28(6):1360–1370, 2012.

63. Yue Zhao, Rong Xiong, Li Fang, and Xiaohe Dai. Generating a style-adaptive trajectory from multiple demonstrations. *Int. J. Adv. Robot. Syst.*, 11(7):103, 2014.

64. Emre Ugur and Hakan Girgin. Compliant parametric dynamic movement primitives. *Robotica*, 38(3):457–474, 2020.

65. Yan Wu, Ruohan Wang, Luis D'Haro, Rafael Banchs, and Keng Peng Tee. Multi-modal robot apprenticeship: Imitation learning using linearly decayed DMP+ in a human-robot dialogue system. 07 2018.

66. Aleksandar Mitrevski, Abhishek Padalkar, Nguyen Minh, and Paul Plöger. "lucy, take the noodle box!": Domestic object manipulation using movement primitives and whole body motion. 07 2019.

67. Jian Fu, ChaoQi Wang, JinYu Du, and Fan Luo. *Concurrent Probabilistic Motion Primitives for Obstacle Avoidance and Human-Robot Collaboration*, pages 701–714. 08 2019.

68. Adam Coates, Pieter Abbeel, and Andrew Y Ng. Learning for control from multiple demonstrations. In *Proc. IEEE Int. Conf. Mach. Learning*, pages 144–151, 2008.

69. Sylvain Calinon and Aude Billard. Statistical learning by imitation of competing constraints in joint space and task space. *Adv. Robot.*, 23(15):2059–2076, 2009.

70. Sylvain Calinon. Robot learning with task-parameterized generative models. In *Robotics Research*, pages 111–126. Springer, 2018.

71. Leonel Rozo, Pablo Jiménez, and Carme Torras. A robot learning from demonstration framework to perform force-based manipulation tasks. *Intel. Serv. Robot.*, 6(1):33–51, 2013.

72. Sylvain Calinon, Florent Guenter, and Aude Billard. On learning, representing, and generalizing a task in a humanoid robot. *IEEE Trans. Syst., Man, Cybern. B, Cybern.*, 37(2):286–298, 2007.

73. Sylvain Calinon, Zhibin Li, Tohid Alizadeh, Nikos G Tsagarakis, and Darwin G Caldwell. Statistical dynamical systems for skills acquisition in humanoids. In *Proc. IEEE-RAS Int. Conf. Humanoid Robots*, pages 323–329, 2012.

74. Christopher G Atkeson, Andrew W Moore, and Stefan Schaal. Locally weighted learning for control. In *Lazy learning*, pages 75–113. Springer, 1997.

75. Sethu Vijayakumar, Aaron D'souza, and Stefan Schaal. Incremental online learning in high dimensions. *Neural Comput.*, 17(12):2602–2634, 2005.

76. Mantas Lukoševičius and Herbert Jaeger. Reservoir computing approaches to recurrent neural network training. *Comput. Sci. Rev.*, 3(3):127–149, 2009.

77. Zhaojie Ju and Honghai Liu. Fuzzy Gaussian mixture models. *Pattern Recognition*, 45(3):1146–1158, 2012.

78. Baibo Zhang, Changshui Zhang, and Xing Yi. Active curve axis Gaussian mixture models. *Pattern recognition*, 38(12):2351–2362, 2005.

79. Freek Stulp, Jonas Buchli, Alice Ellmer, Michael Mistry, Evangelos A Theodorou, and Stefan Schaal. Model-free reinforcement learning of impedance control in stochastic environments. *IEEE Transactions on Autonomous Mental Development*, 4(4):330–341, 2012.

80. Andrej Gams, Bojan Nemec, Auke Jan Ijspeert, and Aleš Ude. Coupling movement primitives: Interaction with the environment and bimanual tasks. *IEEE Transactions on Robotics*, 30(4):816–830, 2014.

81. Martin Tykal, Alberto Montebelli, and Ville Kyrki. Incrementally assisted kinesthetic teaching for programming by demonstration. In *The Eleventh ACM/IEEE International Conference on Human Robot Interaction*, pages 205–212. IEEE Press, 2016.

82. Bojan Nemec, Nejc Likar, Andrej Gams, and Aleš Ude. Human robot cooperation with compliance adaptation along the motion trajectory. *Autonomous Robots*, pages 1–13, 2017.

83. Scott Niekum, Sarah Osentoski, George Konidaris, and Andrew G. Barto. Learning and generalization of complex tasks from unstructured demonstrations. In *IEEE/RSJ International Conference on Intelligent Robots and Systems*, pages 5239–5246, 2012.

84. Emily B Fox, Michael C Hughes, Erik B Sudderth, Michael I Jordan, et al. Joint modeling of multiple time series via the beta process with application to motion capture segmentation. *The Annals of Applied Statistics*, 8(3):1281–1313, 2014.

85. Scott Niekum, Sarah Osentoski, George Konidaris, Sachin Chitta, Bhaskara Marthi, and Andrew G. Barto. Learning grounded finite-state representations from unstructured demonstrations. *International Journal of Robotics Research*, 34(2):131–157, 2015.

86. Chenguang Yang, Chao Zeng, Yang Cong, Ning Wang, and Min Wang. A learning framework of adaptive manipulative skills from human to robot. *IEEE Transactions on Industrial Informatics*, 15(2):1153–1161, 2018.

87. Toshio Tsuji and Yoshiyuki Tanaka. Tracking control properties of human-robotic systems based on impedance control. *IEEE Transactions on systems, man, and cybernetics-Part A: Systems and Humans*, 35(4):523–535, 2005.

88. Christian Ott, Ranjan Mukherjee, and Yoshihiko Nakamura. Unified impedance and admittance control. In *Robotics and Automation (ICRA), 2010 IEEE International Conference on*, pages 554–561. IEEE, 2010.

89. Aliasgar Morbi, Mojtaba Ahmadi, Adrian DC Chan, and Robert G Langlois. Stability-guaranteed assist-as-needed controller for powered orthoses. *IEEE Trans. Contr. Sys. Techn.*, 22(2):745–752, 2014.

90. Brenna D Argall and Aude G Billard. A survey of tactile human–robot interactions. *Robotics and autonomous systems*, 58(10):1159–1176, 2010.

91. Isura Ranatunga, Frank L Lewis, Dan O Popa, and Shaikh M Tousif. Adaptive admittance control for human-robot interaction using model reference design and adaptive inverse filtering. *IEEE Trans. Contr. Sys. Techn.*, 25(1):278–285, 2017.

92. Ning Wang, Chuize Chen, and Chenguang Yang. A robot learning framework based on adaptive admittance control and generalizable motion modeling with neural network controller. *Neurocomputing*, 2019.

93. Yongping Pan and Haoyong Yu. Composite learning from adaptive dynamic surface control. *IEEE Transactions on Automatic Control*, 61(9):2603–2609, Sept 2016.

94. Ning Sun, Yiming Wu, Yongchun Fang, and He Chen. Nonlinear antiswing control for crane systems with double-pendulum swing effects and uncertain parameters: Design and experiments. *IEEE Transactions on Automation Science and Engineering*, PP(99):1–10, 2017.

95. Yongping Pan, Tairen Sun, and Haoyong Yu. Composite adaptive dynamic surface control using online recorded data. *International Journal of Robust and Nonlinear Control*, 26(18):3921–3936, 2016.

96. Fanny Ficuciello, Luigi Villani, and Bruno Siciliano. Variable impedance control of redundant manipulators for intuitive human–robot physical interaction. *IEEE Transactions on Robotics*, 31(4):850–863, 2015.

97. Zhijun Li, Zhicong Huang, Wei He, and Chun-Yi Su. Adaptive impedance control for an upper limb robotic exoskeleton using biological signals. *IEEE Trans. Ind. Electron.*, 64(2):1664–1674, 2017.

98. Zhaojie Ju, Gaoxiang Ouyang, Marzena Wilamowska-Korsak, and Honghai Liu. Surface EMG based hand manipulation identification via nonlinear feature extraction and classification. *IEEE Sensors Journal*, 13(9):3302–3311, Sept 2013.

99. Arash Ajoudani, Nikos Tsagarakis, and Antonio Bicchi. Tele-impedance: Teleoperation with impedance regulation using a body–machine interface. *The International Journal of Robotics Research*, 31(13):1642–1656, 2012.

100. Chenguang Yang, Chao Zeng, Peidong Liang, Zhijun Li, Ruifeng Li, and Chun-Yi Su. Interface design of a physical human–robot interaction system for human impedance adaptive skill transfer. *IEEE Transactions on Automation Science and Engineering*, 15(1):329–340, 2018.

101. Mattia Racca, Joni Pajarinen, Alberto Montebelli, and Ville Kyrki. Learning in-contact control strategies from demonstration. In *2016 IEEE/RSJ International Conference on Intelligent Robots and Systems (IROS)*, pages 688–695, Oct 2016.

102. Abusabah IA Ahmed, Hong Cheng, Huaping Liu, Xichuan Lin, and Mary Juma Atieno. Interaction force convex reduction for smooth gait transitions on human-power augmentation lower exoskeletons. In *International Conference on Cognitive Systems and Signal Processing*, pages 398–407. Springer, 2016.

103. Andrea Cirillo, Fanny Ficuciello, Ciro Natale, Salvatore Pirozzi, and Luigi Villani. A conformable force/tactile skin for physical human–robot interaction. *IEEE Robotics and Automation Letters*, 1(1):41–48, 2016.

104. Huaping Liu, Fuchun Sun, Bin Fang, and Fei Long. Material identification using tactile perception: A semantics-regularized dictionary learning method. *IEEE/ASME Transactions on Mechatronics*, 2017.

105. Peidong Liang, Chenguang Yang, Zhijun Li, and Ruifeng Li. Writing skills transfer from human to robot using stiffness extracted from sEMG. In *2015 IEEE International Conference on Cyber Technology in Automation, Control, and Intelligent Systems (CYBER)*, pages 19–24, June 2015.

106. Fanny Ficuciello, Raffaella Carloni, Ludo C Visser, and Stefano Stramigioli. Port-hamiltonian modeling for soft-finger manipulation. In *Proc. IEEE/RSJ Int. Conf. Intell. Robots Syst.*, pages 4281–4286, 2010.

107. Chenguang Yang, Xingjian Wang, Zhijun Li, Yanan Li, and Chun-Yi Su. Teleoperation control based on combination of wave variable and neural networks. *IEEE Trans. Syst., Man, Cybern. A, Syst.*, 47(8):2125–2136, 2017.

108. Chenguang Yang, Tao Teng, Bin Xu, Zhijun Li, Jing Na, and Chun-Yi Su. Global adaptive tracking control of robot manipulators using neural networks with finite-time learning convergence. *Int. J. Control Autom. Syst.*, 15(4):1916–1924, 2017.

109. Long Cheng, Zeng-Guang Hou, Min Tan, and Wen-Jun Zhang. Tracking control of a closed-chain five-bar robot with two degrees of freedom by integration of an approximation-based approach and mechanical design. *IEEE Trans. Syst., Man, Cybern. B, Cybern.*, 42(5):1470–1479, 2012.

110. Fan Ke, Zhijun Li, Hanzhen Xiao, and Xuebo Zhang. Visual servoing of constrained mobile robots based on model predictive control. *IEEE Trans. Syst., Man, Cybern. A, Syst.*, 47(7):1428–1438, 2016.

111. Zhaoyong Mao and Fuliang Zhao. Structure optimization of a vibration-suppression device for underwater moored platforms using CFD and neural network. *Complexity*, 2017:21, 2017.

112. Chenguang Yang, Xinyu Wang, Long Cheng, and Hongbin Ma. Neural-learning-based telerobot control with guaranteed performance. *IEEE Trans. on Cybern.*, 47(10):3148–3159, 2017.

113. Chih-Min Lin and Ya-Fu Peng. Adaptive CMAC-based supervisory control for uncertain nonlinear systems. *IEEE Trans. Syst., Man, Cybern. B, Cybern.*, 34(2):1248–1260, 2004.

114. Sesh Commuri, Sarangapani Jagannathan, and Frank L Lewis. CMAC neural network control of robot manipulators. *J. Robot. Syst.*, 14(6):465–482, 1997.

115. James S Albus. A new approach to manipulator control: The cerebellar model articulation controller (CMAC). *Trans. ASME: J. Dyna. Syst., Measure., Control*, 97(3):220–227, 1975.

116. Murat Darka. The control of a manipulator using cerebellar model articulation controllers. Master's thesis, İzmir Institute of Technology, 2003.

117. Ning Wang Zhaojie Ju Jian Fu Yang Chenguang, Chuize Chen and Min Wang. Biologically inspired motion modeling and neural control for robot learning from demonstrations. *IEEE Transactions on Cognitive and Developmental Systems*, 11(2):281–291, 2019.

118. Zhijun Li, Jun Deng, Renquan Lu, Yong Xu, Jianjun Bai, and Chun-Yi Su. Trajectory-tracking control of mobile robot systems incorporating neural-dynamic optimized model predictive approach. *IEEE Trans. Syst., Man, Cybern. A, Syst.*, 46(6):740–749, 2016.

119. Chenguang Yang, Chuize Chen, Ning Wang, Zhaojie Ju, Jian Fu, and Min Wang. Biologically inspired motion modeling and neural control for robot learning from demonstrations. *IEEE Transactions on Cognitive and Developmental Systems*, 11(2):281–291, 2018.

2 Robot Platforms and Software Systems

This chapter mainly introduces the relevant robot hardware and software systems. Robot hardware systems include Baxter robot, Nao robot, KUKA LBR iiwa robot, Kinect camera, MYO Armband, and Leap Motion. Robot software systems include MATLAB Robotics Toolbox, CoppeliaSim, and Gazebo.

2.1 BAXTER ROBOT

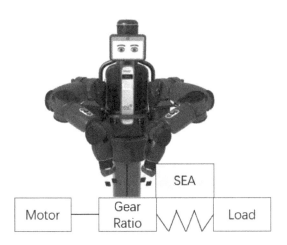

FIGURE 2.1 Baxter research robot profile: schematic SEAs of Baxter robot [1].

The Baxter robot, which is built by Rethink Robotics in the United States, is a three-foot-tall, dual-arm robot. It comprises one torso, one 2-Degrees of Freedom (DOF) head, and two 7-DOF arms, which are shoulder joint, elbow joint, and wrist joint, respectively. The resolution for the joint sensors is 14 bits with 360° (0.022° per tick resolution), while the maximum joint torques that can be applied to the joints are 50 N · m (the first four joints) and 15 N · m (the last three joints). There are also coordinated cameras, torque sensors, position sensors, force sensors, encoders, and sonar with the Baxter robot. Researchers can directly program Baxter using open-source software, such as a standard Robot Operating System (ROS) interface. ROS is an open-source robotic system with libraries, module devices, and correspondences. Seven Serial Elastic Actuators (SEAs) (illustrated in Fig. 2.1) drive all joints of the Baxter robot arm, which gives passive consistency to minimize the constrain of any effect or contact.

Commonly, people program the Baxter robot using ROS through Baxter SDK running on the Ubuntu LTS. It improves the task of displaying and programming on various types of automated platforms. The software development kit of the Baxter robot provides a variety of functionalities to get access to the real-time state of the robot, and it is to implement various control strategies such as position control, velocity control, and torque control.

As an innovative, intelligent, and collaborative robot, Baxter robot has very high adaptability to the environment, which could respond to the changes from the external environment with an interactive safety. Therefore, Baxter robot is an ideal alternative to manpower outsourcing and fixed work automation. With its unique features and benefits, Baxter enables manufacturers to create cost-effective solutions when handling small batches, multivariety production jobs, freeing the hands of technical staff. Nowadays, there are many industrial-leading companies worldwide applying Baxter to their production and have thus obtained a huge commercial competitive advantage [2, 3].

2.2 NAO ROBOT

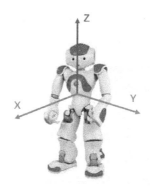

FIGURE 2.2 Nao robot.

Nao robot (shown in Fig. 2.2), produced by the Aldebaran-Robotics in France, is a humanoid robot with 5 DOF joints.

The Nao robot supports multiple sensors, such as cameras, microphones, loudspeakers, chest sonar sensors, movement motors on the neck, hands, and feet, three color LEDs (in red, green and blue) on the eyes, and the head and feet tactile sensors. There are two camera lenses on the NAO, one on the head and the other on the chin. The upper camera views the front, while the lower camera views the foot. The vision system can be used to recognize markers, face recognition, object recognition, image recording, etc. Nao has four microphones: one for each ear, one in front of the head, and one behind the head. These four microphones can be used for simple recording and recognition of the position of the acoustic source. There are two loudspeakers for each ear. Speakers can be used to play music and read text entered by the user. Besides, Nao can also be connected to a wireless network for remote control. Therefore, Nao can communicate with human beings and perceive the environment.

NAO has a body mass index (BMI) of about 13.5 kg/m^2. Hence, compared to other robots with the same height, Nao is quite light. Nao has a total of 25 DOF, wherein 11 joints are for the legs and pelvis, and the rest are for the trunks, arms, and head. In addition, each arm is supported by a 2 DOF shoulder, a 2 DOF elbow, a 1 DOF wrist, and a 1 DOF hand gripper. The head is able to rotate on both yaw and pitch axes [4].

2.3 KUKA LBR IIWA ROBOT

FIGURE 2.3 KUKA LBR iiwa robot.

KUKA LBR iiwa robot (with human-machine collaboration capabilities) is the first mass production of sensitive robots and has been mainly used in the industrial fields. It is able to achieve direct cooperation between humans and robots to complete tasks of high flexible requirements.

The KUKA LBR iiwa robot has an advanced precision robot arm with 7 DOFs. A series elastic element, which connects the motor and harmonic gear reducer with rigid links, is used to drive each joint. Torque sensors and position encoders exist in each of robot's 7 revolute joints. Therefore, motor position and joint or series elastic torques can be sensed, which enable robots to have quick reactions. Its high-performance servo control, which is able to detect contours quickly under force control, makes robots sensitive.

The arm can be programmed via Workbench, which is a standard KUKA modifying platform employing KUKA robot language (KRL) and Java. The KUKA LBR is controlled by the KUKA SmartPad [5, 6], as shown in Fig. 2.3.

2.4   KINECT CAMERA

FIGURE 2.4   Kinect camera.

As a cost effective RGB-D sensor, Kinect is broadly used to capture the point cloud (shown in Fig. 2.4). It is equipped with an RGB color camera, an infrared ray emitter, and a receiver, which enable it to capture the depth information compared to the ordinary camera.

The Kinect camera generates raw data in the form of an RGB image and a depth image. Raw color, depth, and IR images can be captured from the device using the Kinect 2.0 SDK (Software Development Kit) developed by Microsoft. Skeletal tracking can also be achieved with the use of the Kinect 2.0.

Since the point specified by the operator on the Kinect color image is two-dimensional, the Kinect depth space image is required to determine the depth of this point. Both the color image and depth image are transformed into a common frame, the origin of which is located at the center of the depth camera, called Camera space; the coordinate system of this space follows a right-handed convention [7, 4].

2.5   MYO ARMBAND

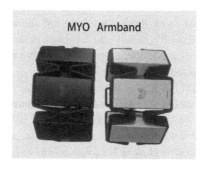

FIGURE 2.5   MYO Armband.

MYO is the brand name of an armband manufactured by Thalmic Labs (as shown in Fig. 2.5). The MYO Armband is a gesture recognition device worn on the forearm and enables the user to control interact with the environment wirelessly using various hand motions.

MYO Armband has 8 built-in sEMG sensors (200 Hz) and one IMU sensor with 9 axes (3-axes geomagnetism, 3-axes angular velocity, and 3-axes acceleration). Hence the hand posture and arm motion can be detected. When the user wears MYO Armband on the forearm and performs different gestures, the muscles in the forearm emit different sEMG signals, which are captured by the built-in high sensitivity sensor of the MYO Armband and processed by the embedded algorithm, so as to recognize different gestures and send gesture instructions to the host computer via Bluetooth.

Because every user has distinctive muscle size, type of skin, and so on, the sensors can create information by electrical driving forces from arm muscles of operators. Critically, the calibrating process is necessary for every user before using the MYO Armbands. Hence, the MYO Armband can identify the motions and gestures of human limbs in a more accurate way.

MYO Armband's application is very extensive, and it can be used for mobile phone and personal computer application process control, such as music playback, PPT playback, and game control. Besides, it can be used for smart cars and aircraft control. It can also be used for medical and health, although most of the application in medical health remains at the level of scientific research, such as deaf-mute sign language recognition, rehabilitation of stroke patients, and prosthetic hand control of amputees [2, 8].

2.6 LEAP MOTION

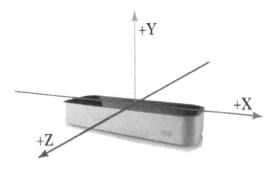

FIGURE 2.6   The coordinate system of Leap Motion.

Leap Motion is a somatosensory controller released by Magic Leap Company. The device is specifically designed to detect and track human hand motion, and to generate a rich set of data about various features from the same hand gestures.

As a miniaturized USB device, it consists of a binocular camera and three infrared LEDs. Leap Motion uses infrared LEDs to follow the trail of the movements of hand and fingers. With this method, we can input hand gesture into an application without other devices such as a mouse or keyboard. In this way, we have the opportunity to create a system to realize gestures recognition in 3D cases.

The origin of the Leap Motions coordinate system is defined at the top-center of the device; specifically, the palms position will be identified as [0, 0, 0], while the user's palm is on the top-center of Leap Motion. Using a right-hand coordinate system, the coordinate system of the Leap Motion controller can be shown as Fig. 2.6 [9]. The XZ axis forms a horizontal plane; the Y-axis is perpendicular; the X-axis is in the direction of the long side of the LM. For Y-axis, the up direction is positive, and the Z-axis is positive in the screen direction.

To make sure Leap Motion works correctly, the hands should be completely horizontal relative to the sensor, and the angle between them should not reach 90°. The sensors will identify each finger's angle, so it also can identify the hand gestures based on identifying the finger's combinations. Compared to the Kinect, the Leap Motion is aiming for hand gesture recognition. We can gain the position of the fingertips and the hand orientation after calculation. In addition, there are also Leap Motion controller (LMC) output depth data frames, including finger position, hand position, rotation, etc. Leap Motion captures and analyzes the data and sends it to the model when the simulation starts, making the model to be able to track the movement of the hand [10]. Besides, the Leap Motion is also used in combination with Head-mounted displays such as Oculus Rift, to obtain a better human-computer interaction performance.

2.7   OCULUS RIFT DK 2

FIGURE 2.7   Oculus Rift DK 2.

Oculus Rift is a virtual reality head-mounted display developed by Oculus VR. Oculus began shipping Development Kit 2 (DK2, as shown in Fig. 2.7) in July 2014. An array of IR LEDs and a custom camera that ships with the DK2 achieve positional tracking. The camera that ships with the device is customized by Oculus VR.

Because of the camera and IR LED array, the device can track the position with submillimeter accuracy. Compared with DK1, it has a higher resolution of 960×180 per eye, and its lenses are larger and clearer. Because of the low persistence, the image of DK2 is significantly clearer when moving head. In addition, it has the features of detachable cable and the omission of the need for the external control box.

Oculus Rift DK 2 is powered directly over USB. A 10-foot long (detachable) cable leads from the DK2 and ends with a USB and HDMI interface. The power button has been moved to the HMD itself, and a powered USB port is on the top of the device which will come in very handy for connecting peripherals.

Oculus Rift DK 2 has a wide range of application scenarios. It can be used in games, social interaction, industrial and professional, watching movies and videos, education, etc.

2.8 MATLAB ROBOTICS TOOLBOX

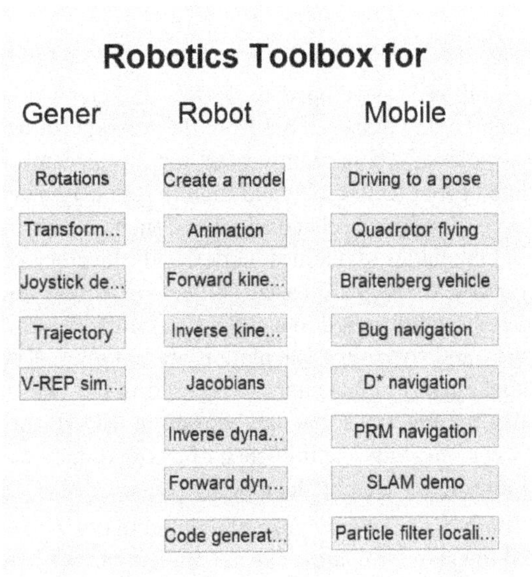

Robotics Toolbox for

Gener	Robot	Mobile
Rotations	Create a model	Driving to a pose
Transform...	Animation	Quadrotor flying
Joystick de...	Forward kine...	Braitenberg vehicle
Trajectory	Inverse kine...	Bug navigation
V-REP sim...	Jacobians	D* navigation
	Inverse dyna...	PRM navigation
	Forward dyn...	SLAM demo
	Code generat...	Particle filter locali...

FIGURE 2.8 Start interface of Robotics Toolbox.

MATLAB is a kind of commercial mathematics software developed by Math-Works company in America. It can be used in advanced technology computing and interactive environment of linear algebra computing, graphics, and dynamic simulation. Now MATLAB has been widely used in university teaching and scientific research. The core functions of MATLAB can be extended through the application-specific toolbox under various commercial or open-source licenses. Vector and matrix are the basic data types of MATLAB; they are very suitable for solving the related problems of robotics.

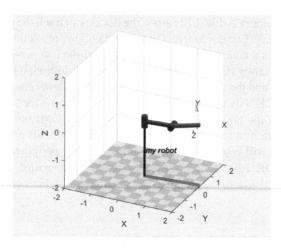

FIGURE 2.9   Model of Robotics Toolbox. An example of a robot model of Robotics Toolbox.

The Robotics Toolbox is developed by Professor Peter Corke during his PhD study at first and can be downloaded free from the website (https://petercorke.com/toolboxes/robotics-toolbox/). It is mainly used for the research and simulation of traditional articulated robot technology and mobile robot. It provides a set of functions to support basic algorithms related to robots, such as directions representation in 3D coordinates, kinematics, dynamics model, and trajectory generation. Most of the examples in robot textbooks are based on two-link robots because the analysis of two-link robots is straightforward to handle. But for the most widely used 6-DOF robot, its kinematics and dynamics calculation are complex. The functions of the robotics toolbox make it very convenient to be used in the two-link robot. In addition, it also contains the functions that can be used in 6-DOF (or more) robots. For example, it is very easy to study the influence of payload mass on the inertia matrix or the change of joint inertia seen by the motor.

The Robotics Toolbox uses a very general method to describe the kinematics and dynamics model of the serial link arm. The parameters of these models are encapsulated in MATLAB objects, so users can use these robot objects to create a variety of series manipulators, and the simulation example makes the abstract learning of robot much more intuitive.

When using the Robotics Toolbox, each link is represented by Link Object. The attributes of each Link Object include Standard or Modified Denavit-Hartenberg parameters, joint and motor inertia values, friction and gear ratio, etc. Several Link Objects make up the robot object, on which the related problems such as forward and reverse kinematics, forward and reverse dynamics can be calculated. The typical models in the toolbox include the classic PUMA560 robot, KUKA robot, and Baxter robot. The start interface of Robotics Toolbox is shown in Fig. 2.8, and a simple robot model of Robotics Toolbox is shown in Fig. 2.9.

SIMULINK is a matching product of MATLAB, which provides dynamic system simulation based on block diagram modeling language. In the Robotics Toolbox, the encapsulation module for toolbox function can describe the nonlinear robot system in the form of block diagrams, which enables users to study the closed-loop performance of the robot control system designed by themselves.

In addition, the Robotics Toolbox also provides conversion tools for dealing with different data types. For example, quaternion (representing three-dimensional position and direction) can use homogeneous transformation through conversion tools to complete the corresponding transformation conveniently and quickly.

The advantages of Robotics Toolbox are:

- The example program in the Robotics Toolbox is very intuitive and easy for users to understand.
- Different implementation methods are provided for the same algorithm, which is helpful for comparative analysis of different situations.
- The Robotics Toolbox provides the source code. When users use it, they can read the source files deeply, to have a better understanding of the use and development of the toolbox.

2.9 COPPELIASIM

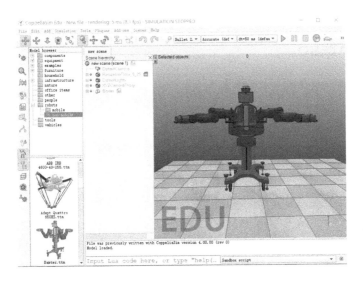

FIGURE 2.10   The Baxter robot in CoppeliaSim.

CoppeliaSim (formerly V-rep, as shown in Fig. 2.10) is an open-source robot simulation software, which can create, compose, and simulate virtual robots. Because CoppeliaSim has many kinds of functions and features, rich application programming interfaces, it is also known as the "Swiss Army Knife" in the robot simulator.

As a robot simulation software, CoppeliaSim can be used for rapid algorithm development, factory automation simulation, rapid prototype design and verification, robot-related teaching, remote monitoring, product safety detection, and other tasks. CoppeliaSim has the following main features:

- CoppeliaSim supports different platforms: Windows, MacOS, and Linux.
- Based on a distributed control architecture, it uses an integrated development environment. Therefore, each object/model can be controlled independently through embedded scripts, plug-ins, add-ons, ROS nodes, remote client application programming interfaces, or customized solutions.
- The controller can be written in multiple programming languages, such as C/C++, Python, Java, Lua, Matlab, Octave, or Urbi.
- CoppeliaSim has four physical engines: Bullet Physics, ODE, Newton, and Vortex Dynamics.
- CoppeliaSim has five main computing modules: Inverse/Forward Kinematics, Minimum Distance Calculation, Collision Detection, Path/Motion Planning, and Dynamics/Physics.
- CoppeliaSim supports customizable particles, which can be used to simulate air, water jets, jet engines, propellers, etc.
- CoppeliaSim has proximity sensor simulation and vision sensor simulation.
- Simulation data is recordable and visible and can be easily imported and exported.

2.10　GAZEBO

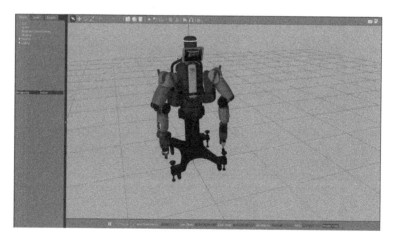

FIGURE 2.11　Gazebo.

Gazebo (as shown in Fig. 2.11) is a software package for simulating robots, which can accurately and effectively simulate a certain number of robots in an indoor or outdoor environment. Gazebo can support highly realistic physical simulation. In addition, Gazebo provides a series of sensors for users and program interfaces.

Gazebo is the default robot simulation software used in the ROS system. Gazebo and ROS are separate projects, that is, their developers are different, but there is a software package related to Gazebo in the official repository of ROS (ros-indigo-gazebo-ros), which is maintained by the gazebo developers themselves. It contains plug-ins of interface ROS and Gazebo. These plug-ins can connect to objects in simulation software scenes and provide simple ROS communication methods, such as topics and services published and subscribed by Gazebo. Encapsulating gazebo as a ROS node also allows it to be easily integrated into the ROS default method running large and complex systems, called boot files.

Although Gazebo and ROS are separate projects, each version of Gazebo development takes into account the adaptation to ROS version so that Gazebo can keep pace with ROS update speed. In general, Gazebo provides an extensive API to access all the functions of any code. Second, Gazebo has integrated some ROS-specific functions, such as services, topic subscription, and publishing. Finally, more importantly, Gazebo and ROS already have a lot of plug-ins and application code/example developed by the community.

In the aspect of scene modeling, Gazebo provides an architectural editor, which is very practical in the design of basic environment and facilities. Second, it provides three simple tracks and shapes: spheres, cubes, and cylinders that can be inserted directly into the scene. Finally, Gazebo has an online model database developed by the community, which is composed of various models and can be accessed directly in Gazebo. This is an advantage of Gazebo in model construction, but the database lacks official maintenance and is relatively unorganized. Users must use external 3D modeling tools, such as Blender or Google SketchUp, to draw models, and then export them into a format supported by Gazebo. Gazebo uses a model format called SDF, which is a file type based on the XML format.

Gazebo has rich functions, among which the typical applications are: robot algorithm testing, robot design, regression analysis of the possible situation of robot reality.

The key features of Gazebo include:

- A large number of physical engines, including ODE, Bullet, Simbody, and DART.
- A rich robot library and environment library.
- Various types of sensors.
- A convenient programmable graphic interface.

REFERENCES

1. Peidong Liang, Chenguang Yang, Zhijun Li, and Ruifeng Li. Writing skills transfer from human to robot using stiffness extracted from sEMG. pages 19–24, 2015.
2. Chunxu Li, Chenguang Yang, Jian Wan, Andy SK Annamalai, and Angelo Cangelosi. Teleoperation control of Baxter robot using Kalman filter-based sensor fusion. *Systems Science & Control Engineering*, 5(1):156–167, 2017.

3. Chenguang Yang, Yiming Jiang, Zhijun Li, Wei He, and Chun-Yi Su. Neural control of bimanual robots with guaranteed global stability and motion precision. *IEEE Transactions on Industrial Informatics*, 13(3):1162–1171, 2016.

4. Chunxu Li, Chenguang Yang, Peidong Liang, Angelo Cangelosi, and Jian Wan. Development of Kinect based teleoperation of Nao robot. In *2016 International Conference on Advanced Robotics and Mechatronics (ICARM)*, pages 133–138. IEEE, 2016.

5. Chunxu Li, Chenguang Yang, Zhaojie Ju, and Andy SK Annamalai. An enhanced teaching interface for a robot using DMP and GMR. *International journal of intelligent robotics and applications*, 2(1):110–121, 2018.

6. Chunxu Li, Chenguang Yang, and Cinzia Giannetti. Segmentation and generalisation for writing skills transfer from humans to robots. *Cognitive Computation and Systems*, 1(1):20–25, 2019.

7. Huifeng Lin, Chenguang Yang, Silu Chen, Ning Wang, Min Wang, and Zhaojie Ju. Structure modelling of the human body using FGMM. In *2017 IEEE International Conference on Cybernetics and Intelligent Systems (CIS) and IEEE Conference on Robotics, Automation and Mechatronics (RAM)*, pages 809–814. IEEE, 2017.

8. Yanbin Xu, Chenguang Yang, Peidong Liang, Lijun Zhao, and Zhijun Li. Development of a hybrid motion capture method using MYO armband with application to teleoperation. In *2016 IEEE International Conference on Mechatronics and Automation*, pages 1179–1184. IEEE, 2016.

9. Congyuan Liang, Chao Liu, Xiaofeng Liu, Long Cheng, and Chenguang Yang. Robot teleoperation system based on mixed reality. In *2019 IEEE 4th International Conference on Advanced Robotics and Mechatronics (ICARM)*, pages 384–389. IEEE, 2019.

10. Zedong Hu, Chenguang Yang, Wei He, Zhijun Li, and Shunzhan He. Modeling and simulation of hand based on opensim and leap motion. In *2017 Chinese Automation Congress (CAC)*, pages 4844–4849. IEEE, 2017.

3 Human-Robot Stiffness Transfer-Based on sEMG Signals

3.1 INTRODUCTION

Research on human motor behavior reveals that human arm can be stabilized mainly by mechanical impedance control during interaction with a dynamic environment [4, 5], which minimizes the interaction force and performance errors. Inspired by this work, biomimetic learning controllers are proposed in Refs. [6, 7, 8, 9] which can be able to simultaneously adapt force, impedance and trajectory in the presence of unknown dynamics. Compared with the traditional robotic controllers, they are "human-like," enabling robots to have some human motor features in an economic perspective, and therefore may have great potentials in compliant human-robot interactions especially in some physically coupling scenarios e.g., rehabilitation or daily tasks.

There exist different kinds of technologies for mapping human mechanical motion, impedance or motor control mechanism to robots due to various body sensors and mathematical models etc. The vision-based model may be a good candidate in transferring human limb movements to the robots [10, 11], but difficult to map force or impedance to robot which may attenuate the transparency among human, robots and environment. Alternatively, sEMG signals may be ideal bio-signals to incorporate human skills and robots [12, 13]. They reflect human muscles activations that represent human joint motion, force, stiffness, etc. [14, 15, 16, 17]. Moreover, sEMG signals are easily accessible and fast adaptive, and the collection of sEMG has little effect on task implementation. sEMG signals are used in different applications (e.g., rehabilitation, exo-skeleton, etc.) or coupled with other sensors (e.g., force, sound and vision [18, 19, 20, 1, 21], etc.). Therefore, sEMG signals are widely used for the robot to obtain human motion information during normal tasks.

As far as humanoid robot manipulator is concerned, it is ideal for transferring human limb dynamic features to the manipulator with elastic actuators because of their geographical similarity [8]. There will be a number of advantages of this human-robot dynamic transfer such as safe, compliant interaction with human and environment with low contact force, low trajectory errors and less time-consuming [22, 23].

In this chapter, sEMG-based human-robot stiffness transfer is discussed. We will explore one human-robot skill transferring method using sEMG-based stiffness transfer. In this method, a dual-arm robot set-up is employed with one robot arm used as a leader to teach motor skills to the other arm as a follower. Force feedback is introduced to enable human demonstrator to feel motion disparity between two arms in a haptic manner, such that the demonstrator would be able to naturally adapt limb stiffness as a response to the applied force. The sEMG signals collected from a MYO armband are utilized to estimate demonstrator's limb stiffness. Instead of estimating absolute human endpoint stiffness [20], differential stiffness estimation is employed to eliminate signal drift and the residual term. Amplitude modulation (AM) envelop of sEMG signals is extracted by squaring and low pass filtering to represent stiffness variations. Stiffness profile, in addition to motion trajectory, is recorded and utilized to realize a much more complete process of skill transfer. An interface is designed combining the above-mentioned techniques, and for natural and easy interaction, a coupling device is purposely designed to couple human and robot physically with hand free so that demonstrator's motion can be naturally transferred to the robot. At last, several experiments are conducted to verify the effectiveness of stiffness transferring in different skill-learning tasks.

3.2 BRIEF INTRODUCTION OF SEMG SIGNALS

Generally, sEMG signals can be processed into two divisions: finite class recognition serials and continuous control reference. The former usually refers to pattern recognition, such as hand posture recognition [24, 25] and such data serials are usually used as switch control signals. In contrast, the latter refers to extract continuous force, stiffness and even motion serials from sEMG signals which especially reflect the variations of human limb kinematics and dynamics during limb movement or pose maintenance. Furthermore, the relationship between sEMG and stiffness, force and motion is approximately linear [14], and thus, bio-controller design tends to be simple in sEMG-based robot control system. In Ref. [26], sEMG signals are processed to extract incremental stiffness to reduce stiffness estimation error and calibration time. Its application is tested via robot anti-disturbance pose maintenance. In Ref. [2], tele-impedance is implemented via continuous stiffness reference, and in Refs. [27] and [28], finger position from sEMG signals are continuously estimated though with rather relatively large error.

sEMG signals represent human muscle co-contractions, and they can be extracted as stiffness regulations and motion pattern, which determines the performance of human-robot interaction and skill transfer. Generally, sEMG envelop detection is a prerequisite for most sEMG post-processing, and sEMG envelop should reflect the features of muscle co-contractions. Moreover, the envelop implementation should be fast and smooth in order to secure a stable and compliant control performance as well as to satisfy robot impedance control requirements.

3.3 CALCULATION OF HUMAN ARM JACOBIAN MATRIX

As shown in Fig. 3.1, human arm is modeled as a 7-DOF arm with 3 DOF on the shoulder, 2 DOF on the elbow and 2 DOF on the wrist. Motion tracking using vision sensors [29] may be significantly affected by occlusion during physical

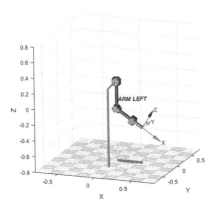

FIGURE 3.1   Human arm kinematics model with 7 DOF.

human-robot interaction. Therefore, we employ filtered IMU readings to measure arm angles, based on the assumption that operator's body and shoulder are stationary.

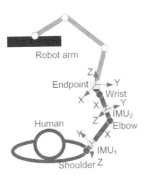

FIGURE 3.2   Human arm Jacobian estimation based on IMUs and inverse kinematics.

As shown in Fig. 3.2, two IMU sensors are worn on the arm, whereas the first one on the upper arm close to the shoulder and the second one on the forearm close to the elbow. They are employed to estimate the first 5 joint angles consisting of 3 shoulder angles and 2 elbow angles, according to the method in Refs. [3, 30]. To calculate the rest 2 wrist angles, we perform simple inverse kinematics using the robot endpoint orientation, because the endpoints of the human arm and robot arm are physically coupled. The joint angle estimation method is summarized in Fig. 3.3.

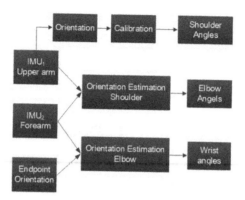

FIGURE 3.3 Human arm Jacobian estimation based on IMU.

3.4 STIFFNESS ESTIMATION

3.4.1 INCREMENTAL STIFFNESS ESTIMATION METHOD

According to the approximated linear relationship between joint force and sEMG signals, human arm endpoint stiffness estimation via sEMG can be calculated by Eq. (3.1).

$$F_h = \begin{bmatrix} A^{ago} & A^{anta} \end{bmatrix} \begin{bmatrix} P_A \\ P_{anta} \end{bmatrix} + \sigma \tag{3.1}$$

where $\begin{bmatrix} A^{ago} & A^{anta} \end{bmatrix} \in R^{3m \times 2n}$, $m = 1, 2$, $m = 1$, omitting the effect of robot endpoint torque; $A^{ago} \in R^{3m \times n}$, $A^{anta} \in R^{3m \times n}$ denote the constant coefficients of agonist and antagonist of muscle co-contractions respectively, moreover, the elements in A^{ago} and A^{anta} have such features: $a_{i,j}^{ago} \geq 0$, $a_{i,j}^{anta} \leq 0$. $F_h \in R^{3m \times 1}$ is the endpoint force generated by muscle co-contractions; $\begin{bmatrix} P_A & P_{anta} \end{bmatrix}^T \in R^{2n \times 1}$ denotes the muscle co-contractions, P_A and P_{anta} denote antagonist and antagonist of muscle co-contractions respectively, which can be represented by filtered sEMG signals, σ is the residual error caused by non-linear factors; n denotes the number of muscle pairs involved.

Therefore, human joint stiffness can be expressed as Eq. (3.2)

$$K_h = \begin{bmatrix} A_h^{ago} & A_h^{anta} \end{bmatrix} \begin{bmatrix} P_A \\ P_{anta} \end{bmatrix} + \sigma' \tag{3.2}$$

where, $K_h \in R^{3m \times 1}$ denotes the endpoint stiffness generated by skeleton muscles involved; the element in $\begin{bmatrix} A_h^{ago} & A_h^{anta} \end{bmatrix}$ is the absolute value of that in $\begin{bmatrix} A^{ago} & A^{anta} \end{bmatrix}$; σ' is the non-linear residual error s and intrinsic stiffness.

The simplified model to estimate human arm incremental stiffness can be expressed as Eq. (3.3) in order to compensate possible non-linear residual items and calibrate human-robot stiffness initialization.

$$\Delta K_h^{t+1} = \sum_{i=1}^{n} |\alpha_i| \cdot \Delta A_t^{anta-i} + \sum_{i=1}^{n} |\beta_i| \cdot \Delta A_t^{ago-i} \tag{3.3}$$

where A_t^{anta-i} is the detected amplitude of sEMG of the ith antagonist muscle at the current time instant t, and A_t^{ago-i} is the amplitude of agonist muscle, $\Delta A_t^{anta-i} = A_t^{anta-i} - A_{t-1}^{anta-i}$, $\Delta A_t^{ago-i} = A_t^{ago-i} - A_{t-1}^{ago-i}$. Note that ΔK_h is the vector of endpoint in Cartesian space, so it needs to be mapped into the robot joint space, which will be discussed in the following section.

3.4.2 STOCHASTIC PERTURBATION METHOD

Three components contribute human arm stiffness: muscle co-contraction activity, arm posture and stretch reflexes [31, 32], which can be synergized in the presentation of Cartesian (endpoint) or joint stiffness. Without considering muscle or posture redundancy, the synergy-based simplified stiffness model can be described as follows [33, 30]

$$K_c = J_h^{+T}(q_h)[K_J - \frac{\partial J_h^{+T}(q_h)f_{ex}}{\partial q_h} - \frac{\partial \tau_g(q_h)}{\partial q_h}]J_h^{+}(q_h) \tag{3.4}$$

where $q_h \in R^7$ is a vector of arm angles of 7 joints angles, $J_h(q_h) \in R^{6 \times 7}$ is arm Jacobian matrix, $K_c \in R^{6 \times 6}$ is the endpoint stiffness, $K_J \in R^{7 \times 7}$ is joint stiffness, $\tau_g(q_h)$ is a vector of gravitational torques and $f_{ex} \in R^6$ is the external force applied onto the endpoint. Obviously, human endpoint stiffness mainly depends on posture which is represented by the Jacobian matrix, as well as joint stiffness and gravity and motion. The arm joint angles q_h of the tutor will be estimated using readings from the gyroscope built in the IMU sensors worn on the tutor. Accordingly, the Jacobian matrix $J(q_h)$ can be calculated using arm length measured beforehand. According to Refs. [14, 33], we assume joint stiffness K_J can be represented as the multiplication of $\bar{K}_J$ and $\alpha(p)$, where $\bar{K}_J$ is an intrinsic constant stiffness identified at the condition of minimal muscle co-contractions, $\alpha(p)$ is an indicator of the coordinated muscle co-contraction, and p is an indicator of muscle activation level to be specified below.

$$\begin{aligned} K_J &= \alpha \bar{K}_J \\ \alpha &= 1 + \frac{\beta_1[1 - e^{-\beta_2 p}]}{[1 + e^{-\beta_2 p}]} \end{aligned} \tag{3.5}$$

where β_1 and β_2 are constant coefficients to be estimated, and p is calculated in the following manner. First, low pass filtering and moving average techniques are used to extract an envelope of the raw sEMG signals from each of the 8 channels using the algorithm shown in Fig. 3.4. The moving average process takes the following equation

$$f(A_t) = \frac{1}{W} \sum_{k=0}^{W-1} EMG(A_{t-k}) \tag{3.6}$$

where $f(A_n)$ is enveloped sEMG amplitude, W is the window size, $EMG(A_k)$ is the sEMG signal amplitude at sample point k, and t is the current sampling time. The absolute value of enveloped sEMG amplitude from each of the $N = 8$ channels is then summed together to yield an indicator of coordinated muscle activation level as defined below

$$p(k) = \sum_{i=1}^{N} |f_i(A_t)| \tag{3.7}$$

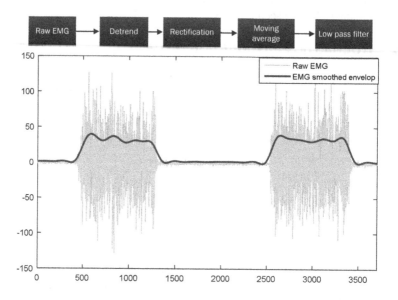

FIGURE 3.4 Filtered sEMG envelope by using low pass filter and moving average method.

The unknown parameters β_1 and β_2, as well as constant intrinsic joint stiffness matrix $\bar{K}_J$ is identified based on the following minimization task

$$\min \left\| \alpha(p)\bar{K}_J - \frac{\partial J_h^{+T}(q_h) f_{ex}}{\partial q_h} - J_h^T(q_h) K_c J_h(q_h)) \right\| \tag{3.8}$$

which is derived from equation (3.4) by omitting the effect of gravity due to the armrest used in the experiment. In order to obtain the endpoint stiffness K_c at different postures, we follow the procedure developed before in Ref. [22] to identify a force and position deformation mapping defined below:

$$\begin{bmatrix} f_x \\ f_y \\ f_z \end{bmatrix} = \begin{bmatrix} G_{xx} & G_{xy} & G_{xz} \\ G_{yx} & G_{yy} & G_{yz} \\ G_{zx} & G_{zy} & G_{zz} \end{bmatrix} \begin{bmatrix} \Delta x \\ \Delta y \\ \Delta z \end{bmatrix} \tag{3.9}$$

where f_x, f_y, and f_z are forces generated by robot arm and applied on arm endpoint along x, y, z-axis, respectively; $G_{*,*}$ is a second-order impedance model; Δx, Δy, Δz are the deformation caused by the applied force.

3.5 INTERFACE DESIGN FOR STIFFNESS TRANSFER

The scheme of the pHRI system is shown in Fig. 3.6. A mechanical module and sEMG arrays are combined as the interface for human impedance adaptive skill transfer [23]. The former is used to physically couple the human tutor hand and the leader arm of Baxter robot. It is also employed to improve the measurement precision of endpoint force by confining wrist joint motion during the calibration process of human arm endpoint stiffness. And the latter is utilized to collect raw sEMG signals from human tutor limb for human hand motion classification and human impedance estimation. Human hand gesture classification is exploited to decode human hand motion patterns and applied in switch-like control such as robot gripper open/close and robot control modes switching. The classification results are generally mapped into robot control switching strategy to fulfil a specific task. Human stiffness estimation in the interface is realized based on the FSE and dimensionality reduction methods, where the main active muscles are chosen in the implementations. Stiffness estimated in an almost real-time manner from the sEMG signals is transferred to the robot through a variable impedance control model. Fig. 3.6 also shows the scenario of interface application in a specific task, where the robot follower arm is guided by the tutor to learn human-like behaviors.

The profile of the mechanical coupling module is shown in Fig. 3.7. The endpoint of the robot and the module are joined together through the flange adaptor. The other side of the module is designed in accordance to human arm profile to connect the tutor's arm. It couples the human arm with the robot but with the human hand free of motion in a specific task. The coupling module can be adapted to different human tutors with the use of the wrist size adapter. The force sensor mounted on the module is used to measure the end-effector force in order to identify the mapping of sEMG-to-stiffness. Once the parameters are determined, the sensor is no longer needed for the following experimental tasks. The inner surface of the coupling module is usually covered with soft sponges such that human tutor will feel comfortable when he/she drives the robot arm to move in the teaching-learning phase.

The functionalities of the designed coupling interface are further depicted in more details, as shown in Fig. 3.5. There are several characteristics of the interface that can be illustrated as follows [23]:

TABLE 3.1
Functionalities of the Components

Components	Functionality	Components	Functionality
User wrist size adapter	Adjusting different users	Baxter robot adapter	Assembling on Baxter robot
Clock-type dial	Adapting decoupling force	Soft sponge	Making users comfortable

First, haptic feedback is introduced into the pHRI system through the designed sEMG-based interface. sEMG signals are detected for both hand gesture recognition and variable stiffness estimation. It allows the human tutor to directly "feel" the

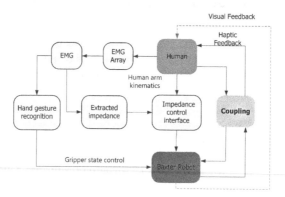

FIGURE 3.5   The functionalities of the designed interface in the pHRI system: the human tutor and the robot are physically coupled through the mechanical module. The haptic and visual feedback is provided to the tutor for skill transfer.

motion difference between the leader arm and the follower arm to be taught in almost real-time, thus enabling the tutor to naturally and intuitively adjust his/her stiffness on-line to adapt to variation of a task and dynamic environment based on the haptic feedback as well as the visual feedback. Therefore, the proposed interface makes skill transfer more natural than that with only visual feedback without the involvement of the haptic feedback mechanism. In this work, haptic feedback is realized through the dual-arm control strategy, which will be illustrated in a later subsection.

Second, safety should be prioritized for the pHRI systems [34]. The disengagement mechanism is designed to guarantee safety in the mechanical coupling module. The friction forces between the flank of base and the four connecting screws can be accordingly increased or reduced by tightening or loosening the screws in different situations such that the tutor's hand is able to safely disengage from the robot in case of a sudden malfunction of the robot. Threshold force, which can be roughly measured by a force sensor, is utilized in the module to secure the safety. A safety test is carried out to test the chosen threshold before an experimental task is performed. It needs to be emphasized that the safety test should be conducted with respect to different human tutors because of the differences in muscle strength. Fig. 3.8 is a typical example of a safety test. In the experiment, pre-tighten forces of the screws are first set by properly adjusting the screwing length of the screws. Human tutor's hand is then connected to the module. The tutor tries to get disengaged from the module by pulling the brace such that the contact force can be obtained. The force will increase when a human tutor intentionally stops movement with the robot leader arm. As the impulse force point is shown in Fig. 3.8, the bracer will automatically get disengaged when the force grows beyond the pre-set threshold. More specifically, once an accident occurs and human hand gets constrained, the human tutor is able to disengage from the robot without getting injured by giving the bracer a stronger

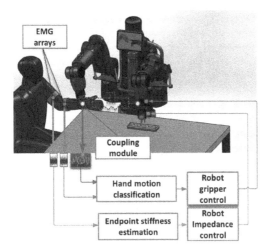

FIGURE 3.6   Overall illustration of the phRI system: the sEMG arrays are used for detecting human tutor's muscle activities. The coupling module is connected to the robot leader arm's endpoint for physical interaction with the tutor.

pull, to make the screws less tighten and get the bracer detached from the main part of the mechanical module.

Moreover, the interface is developed for bimanual leader-follower movements, which enables the follower arm to follow the motion of the leader arm. Therefore, the follower arm movement will not be disturbed by the human tutor, and the tutor's sight will also not be hindered.

3.6   HUMAN-ROBOT STIFFNESS MAPPING

A dual-arm robot platform is used in this work, whereas the leader arm is coupled with the human tutor's hand, and the follower arm is guided to learn the featured human behaviors [23]. The two arms have almost the same kinematic and dynamic parameters. Fig. 3.9 shows the human-in-the-loop robot system. In addition to visual feedback, haptic feedback is developed and integrated into the control strategy, thus enabling the human tutor to subconsciously perceive the performance of the follower arm movement. The haptic feedback also helps the human tutor to naturally adjust his/her arm stiffness.

In the control model, a tracking error-based feedback mechanism is employed to produce haptic feedback. It means that the haptic feedback is based on the motion tracking error between the robot leader and follower arms. The former is driven to move by the human tutor, and the latter follows the instructions. The two arms can be seen as a leader-follower system represented by a virtual spring-damper coupling model. Let D_f and K_f be the feedback damping and stiffness, respectively; $\dot{x}_s$ and x_s be the velocity and position of the follower robot arm endpoint; $\dot{x}_m$ and x_m be the

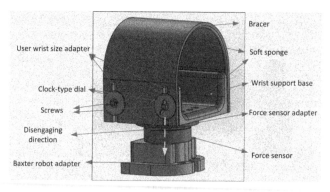

FIGURE 3.7 Profile of the mechanical module of the interface: the design is made purposely for Baxter robot arm and can be easily adapted for other robots. Human tutor can adjust tightness by adjusting the screws. The amplitude of the maximal tightness can be visualized on the clock-type dial. See Table 3.1 for the functionalities of the main components.

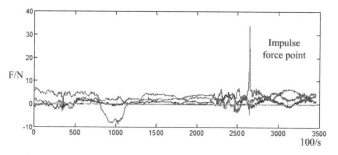

FIGURE 3.8 A typical result of the safety test: the bracer will get detached from the main part of the mechanical module once the force is beyond the pre-set threshold as shown at the impulse force point.

velocity and position of the leader arm endpoint; τ_g is the gravity. Then, the haptic feedback F_f can be calculated as below:

$$F_f = D_f(\dot{x}_s - \dot{x}_m) + K_f(x_s - x_m) + \tau_g \qquad (3.10)$$

which can be then transformed into the joint space as:

$$\tau_f = J_r^T F_f \qquad (3.11)$$

where J_r is the Jacobian matrix of the robot arm. $q_{m,i}$ and $q_{s,i}$ are the ith joint angle of the leader arm and the follower arm, respectively. The damping matrix is set as $D_f = \sigma_f K_f$ with a proper scale factor σ_f. The haptic feedback τ_f reinforces human tutor's haptic feeling of the follower arm's performance to trigger his/her stiffness adaptation. The follower arm is controlled by a PD controller with variable gains such that it is able to track the movement of the leader arm well. The controller is

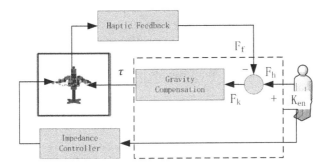

FIGURE 3.9 Schematic of the human-in-the-loop robot system: a virtual spring model is developed for the leader-follower system. A tracking error-based haptic feedback is integrated into the dual-arm control strategy.

illustrated as below:

$$\tau_s = D_s(\dot{q}_m - \dot{q}_s) + K_s(q_m - q_s) \tag{3.12}$$

where $D_s = \sigma_s K_s$ with a proper scale factor σ_s, and variable stiffness K_s, which is obtained according to the following equations:

$$K_{sx} = J_r K_{en} J_r^T \tag{3.13}$$

$$K_s = K_{smin} + \frac{(K_{sx} - K_{smin})^2}{K_{smax} - K_{smin}} \tag{3.14}$$

where K_{smin} is the minimum stiffness of Baxter robot, K_{sx} is the joint stiffness mapped from human arm endpoint, and K_{smax} is the maximum stiffness of Baxter robot.

3.7 STIFFNESS TRANSFER FOR VARIOUS TASKS

3.7.1 COMPARATIVE TESTS FOR LIFTING TASKS

The experimental studies use a dual-arm Baxter robot. Moreover, joint stiffness and damping can be modified under the Baxter robot torque control mode, which is simplified as PD impedance control law in chapter 2. To verify the efficiency of proposed TbD method, human-robot telemanipulation for grasping/lifting/moving an object is implemented with comparative tests under three different set-ups (refer to Table 3.2). The robot left arm serves as the leader arm (tutor) which is coupled with human demonstrator's limb endpoint via a purposely built coupling device shown in Fig. 3.7. The right arm serves as the follower arm (tutee) to follow the leader arm. Two MYO armbands are worn by a human demonstrator. One is worn on the forearm for hand gesture recognition and detection of Brachioradialis muscle co-contractions, and the

TABLE 3.2

Comparative Tests on Baxter Robot and Implementation of Test Steps on Each Test.

Test No.	sEMG	Force feedback	I	II	III	IV	V
Test I	×	×	√	√	√	×	√
Test II	√	×	√	√	√	√	√
Test III	√	√	√	√	√	×	√

other is worn on the upper arm to detect Biceps and Triceps muscle co-contractions that drive elbow motions, as shown in Fig. 3.10:

FIGURE 3.10 MYO armband worn on the upper arm used for sEMG signal measurement. Signals collected from all its 8 channels are utilized to detect elbow flexion/extension.

As shown in Fig. 3.11, human demonstrator teaches the follower robot arm to grasp, lift and move a light object (green) first, and then a heavy object (brown) in three comparative tests as specified in Table 3.2, where × indicates "No" and √ indicated "Yes." Columns 4 to 8 in Table 3.2 show whether a particular test includes one of these tasks: (I) ready to pick up; (II) picking up the light object (0.2Kg); (III) picking up the heavy object (1Kg); (IV) picking the heavy object using visual feedback, and (V) returning to the start position. For the convenience of implementation, only three joints ($Left\,S1$, $Left\,E1$, $Left\,W1$) of Baxter robot arms are used, while the rest four joints are fixed by high gain control method. Human tutor's right arm endpoint is coupled with the Baxter left arm (leader arm) endpoint using coupling device and human tutor's elbow joint rest on an arm-support, and the elbow joint is free to perform lifting in the vertical axis Z and flexion/extension in the horizontal plane (X, Y), while shoulder joint is kept static during these tasks.

The Tbd-based hand gesture recognition algorithm is used on the sEMG signals collected from MYO armband worn on the forearm arm, and it returns an indicator of hand motion, as shown in Fig. 3.12. This will be used to control the gripper of the Baxter robot to grasp or to drop an object.

Test I: grasping-lifting-moving an object with fixed control parameters properly selected to enable Baxter robot to implement tasks with low motion error and proper contact force. In addition, there is no force feedback provided to the human demonstrator. Fig. 3.13 shows the positions and contact forces of the leader arm and follower arm, respectively. It can be seen that the follower arm is able to follow the leader arm when a light load is lifted, while it fails to follow for heavy load.

FIGURE 3.11 Human-robot skill transfer for grasping-lifting-moving an object: human tutor and Baxter left arm which provides force feedback are coupled physically by a purposely designed coupling device shown in Fig. 3.7.

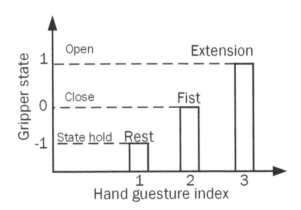

FIGURE 3.12 Hand gesture index mapping for Baxter gripper control.

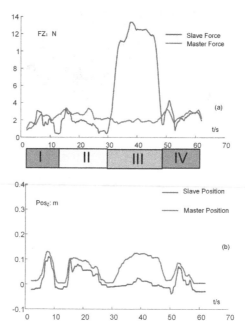

FIGURE 3.13   Human-robot TbD test I: (a) forces of Baxter robot dual-arms. The leader arm force are nearly constant and follower arm position has no effect on leading force; (b) position of Baxter dual-arms. The robot arm position error is dependent on external load. Test is divided into 4 stages shown in Fig. (a): Free motion, light object lifting, heavy object lifting and returning to start position.

Test II: Grasping-lifting-moving an object with sEMG-based variable stiffness transferring. Compared with Test I, follower arm joint stiffness is set in real-time to adapt to payload change according to human demonstrator's response to visual feedback. When the position difference (in a symmetric manner) between leader arm and follower arm increases, the human tutor could increase elbow joint impedance intentionally to increase follower arm's stiffness to lift up a heavy object as shown in Fig. 3.14. To demonstrate the performance of human-operated control, human tutor first completes the same tasks as in Test I. When lifting heavy objects, the human tutor would increase his/her elbow joint stiffness to decrease the position error between two arms. Fig. 3.14 shows the recorded raw sEMG signals (8 channels), the estimated stiffness during tasks, as well as the position and force of each arm of the Baxter robot. The test shows that with visual feedback only, stiffness estimated from human demonstrator would not increase naturally to lift heavy objects. Only when the human demonstrator intentionally stiffens up by co-contraction of muscles, the follower robot arm reduces the tracking error with the leader arm.

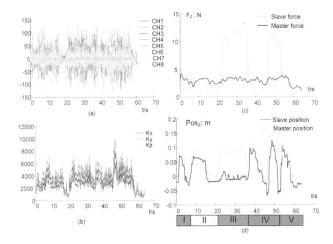

FIGURE 3.14 Human-robot LfD test II: (a) Raw sEMG signals; (b) Estimated stiffness using envelope of the smoothed sEMG signals; (c) Force of leader and follower arms; (d) Position of leader and follower arms.

Test III: Grasping-lifting-moving an object with sEMG-based variable stiffness transferring and force feedback. In this test, human demonstrator would repeat the tasks of Test I and would tune the stiffness when lifting according to the force feedback from the follower arm. Fig. 3.15 clearly shows that the position tracking error reduces immediately and dramatically when lifting the heavy object compared to Tests I and II; because the human demonstrator is able to subconsciously increase stiffness without visual feedback. Therefore, it is more natural and convenient to teach a robot with force feedback. Moreover, the stiffness response detected from sEMG signals is more sensitive so that more efficient TbD in a dynamic environment can be achieved.

3.7.2 WRITING TASKS

The experimental platform used in this experiment is the Baxter robot. The telecommunication system used in this experiment is based on a distributed control system that is Robot Raconteur [35] (Robot Raconteur Version 0.4-testing is used in this experiment) which is illustrated in Fig. 3.16.

Before the writing task, three steps need to be implemented: (A) Calibration, 3D incremental stiffness calibration via sEMG signals is prerequisite for transferring human natural impedance regulation to the robot side. In this experiment, the estimation model is shown in Fig. 3.17. The subject is asked to move in $\pm X$, $\pm Y$, $\pm Z$, and random directions with his/her wrist attached to coupling mechanism in order to constrain its motion as shown in Fig. 3.18. The force exerted $\pm 5N$, $\pm 10N$, $\pm 15N$ is monitored by force presentation interface shown in Fig. 3.18 via visual feedback, so

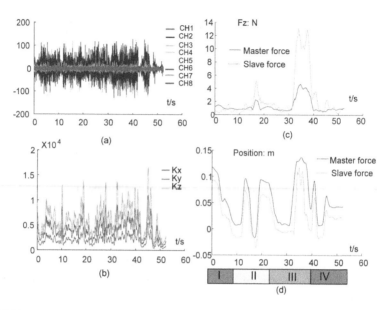

FIGURE 3.15   Human-robot LfD test III: (a) Raw sEMG signals; (b) Estimated stiffness using envelope of the smoothed sEMG signals; (c) Force of leader and follower arms; (d) Position of leader and follower arms.

we can obtain incremental force array $R^{8 \times 1}$ in each coordinate axis, and the corresponding sEMG signals which would be differentiated are recorded synchronously using NI USB6210 (16 inputs 16bit, 250KS/s, and multifunction I/O), 8 channels sEMG Pre-Amplifier with sampling rate 2K Hz. 8 muscles are involved in the endpoint stiffness incremental estimation, that is at least 8×3 trials should be worked out to calculate the coefficient of endpoint stiffness, while in this experiment, 54 trails are implemented, that is, 3 trails are carried out in each direction of $\pm X$, $\pm Y$, $\pm Z$. The least-square method is utilized to improve the precision of estimation. The estimated endpoint stiffness coefficient is $R^{3 \times 8}$ matrix. (B) stiffness mapping from human task space to robot joint space should be under robot stable boundaries and reflect the range of human endpoint stiffness as well. The mapping model is shown in Fig. 3.19. (C) impedance interface design for tests: all the comparative tests are based on impedance design that the human arm endpoint movement will be the reference trajectory and recorded via Baxter position sensors, the mapping stiffness will be the gain of position error in the feedback loop and the gain for velocity is the damping which will be simplified as Eq. 3.16 where 0.2 is selected empirically. So overall experimental components are integrated as a human-robot transferring system which can be illustrated in Fig. 3.19. The impedance interface is a PD controller expressed as Eq. (3.15)

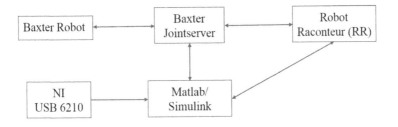

FIGURE 3.16   Telecommunication system design for Baxter.

$$\tau = -k_d(\dot{q} - \dot{q}_d) - k_p(q - q_d) \qquad (3.15)$$
$$K_d = 0.2 * \sqrt{K_q} \qquad (3.16)$$

In the Baxter impedance control mode, human-robot writing skills transfer Fig. 3.20 is implemented as follows. First, human-robot cooperative writing tasks: one subject (right-handed) right wrist is coupled with Baxter arm endpoint using the coupling mechanism; meanwhile, sEMG electrodes are attached to the eight muscles shown in Fig. 3.21. A4 paper is attached on the desk, which can be seen as a plain surface. The subject right hand used Baxter robot to write a simple Chinese word "kou" on the paper like using a pen when the sEMG signals and Baxter robot endpoint position are recorded for the later comparative tests. sEMG signals are processed through incremental stiffness estimation interface. Then three tests are carried out: (i) Baxter robot wrote the same word using the reference position recorded during the human-robot cooperative writing tasks, while the endpoint stiffness is set to [800 N/m, 800 N/m, 800 N/m]; (ii) Baxter repeated the first writing task, but the endpoint stiffness is set to [100 N/m, 100 N/m, 100 N/m] to make it compliant; (iii) Baxter robot wrote the word "kou" with the recorded position while the endpoint stiffness is set to the human arm endpoint stiffness via human-robot stiffness mapping mechanism. Performances can be represented as Fig. 3.26. Figs. 3.22 and 3.23 show the endpoint contact force and position error under high stiffness, low stiffness and variable stiffness extracted from sEMG signals. To further indicate the comparative efficiency of these three writing tasks, correlations of corresponding contact forces are calculated shown in Fig. 3.24.

3.7.3   HUMAN-ROBOT-HUMAN WRITING SKILL TRANSFER

Human-robot-human writing skill transfer is implemented on a Baxter Robot to perform comparative tests under four different setups as specified in Table 3.3. The Baxter robot's left arm is regarded as the mater arm with a pen mounted, while the right arm is regarded as the follower arm grasped by the tutee who is physically guided by the arm. Two MYO sensors are worn on the tutor: one on forearm close to the elbow, and the other on upper arm close shoulder. As illustrated in Fig. 3.25,

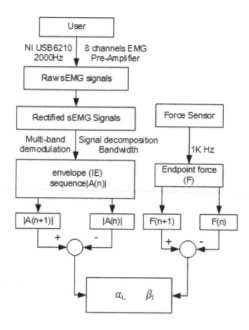

FIGURE 3.17 Incremental stiffness coefficients estimation procedure.

the MYO sensors are used in this work to measure both muscle activation level indicator $\alpha(p)$ and arm posture [30]. In the experiment, the human tutor is supposed to teach tutee to write a Chinese character "Shui," under 4 comparative tests as indicated in Tab. 3.3. The experimental system is shown in Fig. 3.27 (right). The human tutor's right hand holds a pen mounted on the endpoint of Baxter's left arm (the leader arm), and the right arm (the follower arm) physically guides the tutee to write in the three tests.   The experimental results of writing performance at each stage

.

TABLE 3.3
Specification of the Comparative Tests

Test item	sEMG	stiff-force feedback
Test I (soft mode)	×	×
Test II (rigid mode)	×	×
Test III (sEMG mode)	√	×
Test IV (sEMG & haptic feedback)	√	sEMG stiff-force feedback

are illustrated in Fig. 3.28. The processed sEMG signals during different stages are shown in Fig. 3.29. In the teaching stage, the force and stiffness feedback provides a natural perception of tutee's performance to the tutor, as if they are connected. This "connection" in fact disturbs the normal writing performance of the tutor, who must

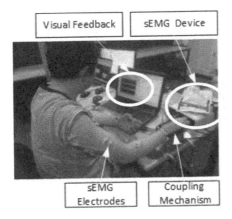

FIGURE 3.18 Stiffness calibration based on force sensors and sEMG signals.

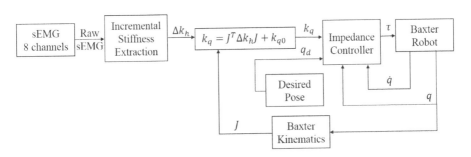

FIGURE 3.19 Stiffness mapping design from human to Baxter robot and interaction based on impedance control.

FIGURE 3.20 Human robot writing skills transfer experimental setup.

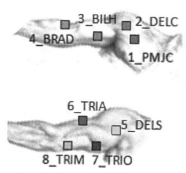

FIGURE 3.21 Muscles, number and positions of electrodes involved in the endpoint stiffness estimation [20].

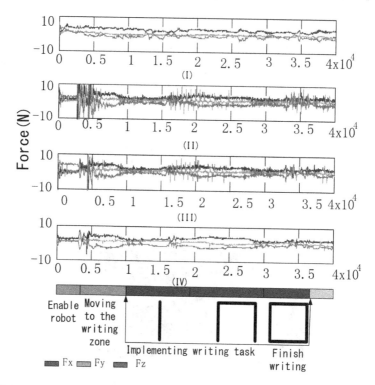

FIGURE 3.22 Comparative writing skills performance: contact force with three different endpoint stiffness scenarios: (I) Reference writing force; (II) writing under constant high stiffness [800 N/m, 800 N/m, 800 N/m]; (III) writing under variable stiffness via sEMG signals. (IV) writing under constant low stiffness [100 N/m, 100 N/m, 100 N/m].

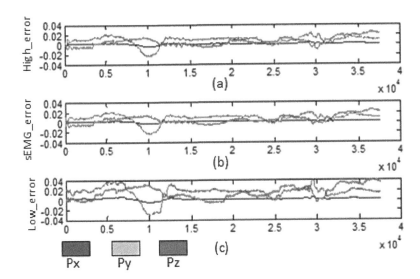

FIGURE 3.23 Comparative writing skills performance: position error with three different endpoint stiffness scenarios: (a) writing under constant high stiffness [800 N/m, 800 N/m, 800 N/m]; (b) writing under variable stiffness via sEMG signals. (c) writing under constant low stiffness [100 N/m, 100 N/m, 100 N/m].

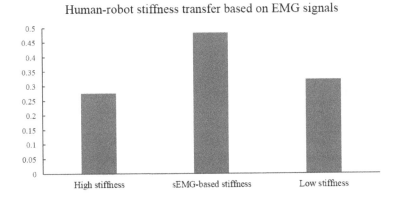

FIGURE 3.24 Correlation in Z direction among reference force, sEMG-based force, high stiff force and low stiff force.

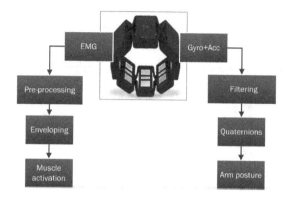

FIGURE 3.25 The structure of signal collection and processing from the MYO sensor.

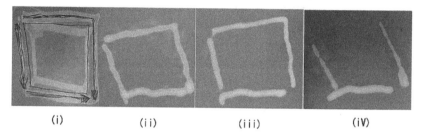

    (i)   (ii)   (iii)   (iV)

FIGURE 3.26 Writing results: (i) Reference writing; (ii) High stiffness writing; (iii) sEMG-based stiffness writing; (iv) Low stiffness writing.

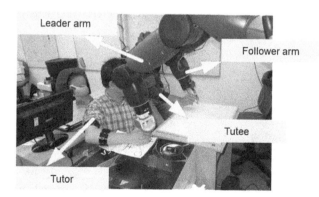

FIGURE 3.27 Setup of human-robot-human writing skill transfer: robot left arm (leader arm) provides force and stiffness feedback to human tutor and captures the motion, which is transferred to the robot right arm (follower arm) which is identical to the left arm in terms of kinematics. Human tutee is physically guided by the right arm, which is controlled with variable stiffness generated according to the human tutor's muscle activations.

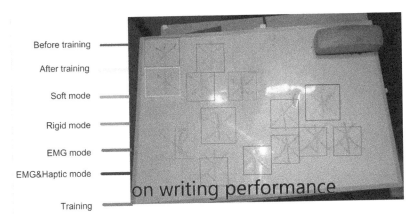

FIGURE 3.28 Experimental results of the writing performance under four control modes.

respond with the adaptation of force and stiffness in order to perform the writing as normal. The response of the tutor is unique according to different characteristics of the tutee's motor behaviors, thus personalized guidance can be provided to the tutee. In this experiment, we evaluate the writing performance in terms of hand trajectory and stiffness adaptation.

One tutor (aged between 20–30) who is good at Chinese handwriting and one tutee without any knowledge of Chinese handwriting participated in the tests under different setups as specified in Table 3.3. Analysis of test results mainly will be performed by comparison of the results under different modes, as shown in Fig.3.30.

(i) Soft mode: Handwriting guidance to tutee is provided by a low gain controller, whereas the control gains are properly chosen to enable the robot performing tasks with low position error as well as contact force. There is no haptic feedback from tutee to tutor such that visual feedback is the only means for a tutor to evaluate the performance of tutee. In terms of both motion tracking error and motion error variance, soft mode perform worst.

(ii) Rigid mode: Handwriting guidance to tutee is provided by a high gain controller with control gains properly selected with guaranteed stability. There is no haptic feedback, and similarly to Test (i), the tutor only observes the tutee's performance using vision. It performs best in terms of motion tracking and tracking variance. However, using a fixed high gain control is brutal and does not provide assistance to the tutor in a personalized manner according to individual tutee's motor performance. As shown in the analysis of forces, there is a low correlation between forces of the leader arm (feedback force) and of the follower arm (driving force). When the tutee is able to actively follow the tutor's motion, the guidance force should be reduced to encourage the self-drive motion. But under Rigid mode, the tutee receives large force constantly.

(iii) sEMG mode: Handwriting guidance to the tutee is provided by a controller

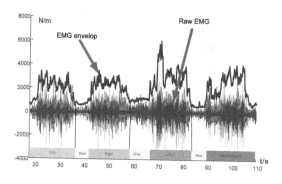

FIGURE 3.29   sEMG processing during the 4 control modes for handwriting.

of variable stiffness, to be set by the muscle activations of the tutor in the absence of any haptic feedback. The tracking performance of tutee has been greatly improved in comparison with Test (i) under soft mode. In addition, there is a significant correlation between force feedback to tutor and the force driving tutee, implied the teaching adapts to the performance of learning. However, the adaptation of muscle activation using only visual feedback is not as natural as in the presence of interactive force. Thus the teaching experience is not guaranteed.

(iv)   sEMG and haptic feedback mode: Handwriting guidance to tutee is provided by a controller of variable stiffness, to be set by the muscle activations of the tutor in the presence of haptic feedback, which provides an intuitive perception of the tutee's performance. Similarly to the sEMG mode in Test (iii), the correlation between driving force on the follower arm for tutee and feedback force on the leader arm on the tutor is obvious. While in terms of tutee's tracking performance as well as tracking variance, it outperforms the results under the sEMG mode without haptic feedback in Test (iii). Haptic feedback makes it much easier for the tutor to adapt muscle activation as a response to the tutee's performance. Thus, teaching experience is much improved, in addition to the improved learning performance.

3.7.4   PLUGGING-IN TASK

Electrical Socket Plugging-in Task: The experimental setup [23] is shown in Fig. 3.31. Human tutor's arm is coupled with the robot leader arm, and the tutor drives the leader arm to move, then the follower arm follows the movement based on the developed dual-arm control method. Baxter robot is expected to complete the three subtasks: (i) grasping controlled by hand gestures; (ii) positioning assisted by human tutor's haptic and visual feedback; (iii) plugging into the socket with motion and stiffness adaptation. For comparison, the plugging-in task is conducted in the following four modes:

I) Default teaching mode: the robot performs the socket plugging-in task in the

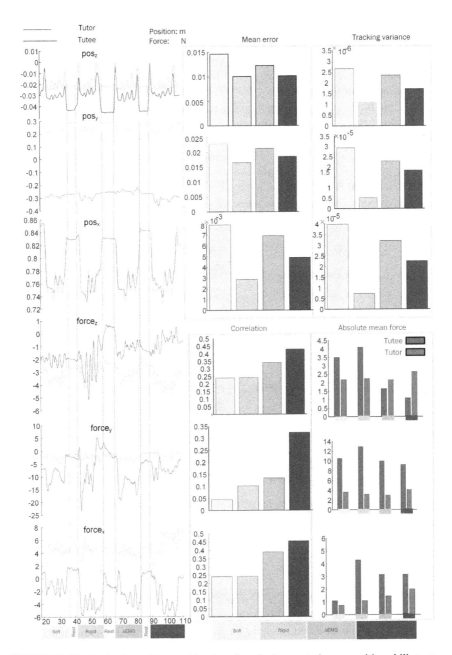

FIGURE 3.30 Evaluation of the teaching interface for human to human writing skills transfer under four control modes.

position control mode provided by the manufacturer. Human tutor performs a plugging-in demonstration, and then the robot replays the motion.

II) Soft mode: human tutor guides the robot to complete the plugging-in task in torque control mode with fixed low gain stiffness. Then the robot replays the motion using the same low gain stiffness.

III) Rigid mode: human tutor guides the robot to complete the plugging-in task in torque control mode with fixed high gain stiffness. Then the robot replays the motion using the same high gain stiffness.

IV) The sEMG-based mode: human tutor guides the robot to complete the plugging-in task with variable stiffness estimated based on the sEMG signals which are extracted from the tutor's arm in the presence of haptic feedback. Then the robot replays the motion with the same adaptive stiffness. The experimental results and

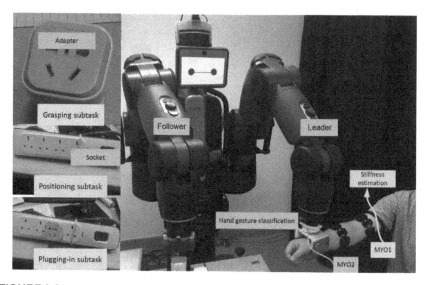

FIGURE 3.31 Experimental setup for the plugging-in task: human tutor's hand is physically coupled with the robot leader arm through the module. The robot follower arm follows the movement of the leader arm to learn the tutor's skills for the task. The sEMG arrays are used to detect the muscle activities of human tutor. MYO1 is for stiffness estimation and MYO2 is for hand gesture classification.

discussions are illustrated as follows: Plugging-in Task in Default Teaching Mode: In this mode, the human tutor drives the robot arm to perform the plugging-in task by grasping the touchpad on the wrist of Baxter's arm. Then the robot replays the movement. Fig. 3.32 (Left Top) shows the position trajectories of human tutor and the robot. It indicates that the robot is able to imitate human tutor's positioning skill well (the maximum position error of 0.1956 cm in the y-axis). However, when the robot replays the plugging-in task, it seems to be difficult to plug the adaptor into the socket. In the position control mode, the desired stiffness cannot be achieved. So the robot positioning cannot be guaranteed in the physical contact between the adapter and the socket. Besides, the movement repeatability of the robot also affects performance. Fig. 3.32 (Left Top) also shows the difference of the contact forces. The human tutor is able to adapt to the task in the teaching phase. While obviously, it is difficult for the robot to imitate such human adaptive skills using the automatic mode during the phase of play-back.

Plugging-in Task in Soft Mode: In this mode, a human tutor performs the plugging-in task with the Baxter robot by driving the leader arm through the interface. The tutor's hand is not constrained but free to transfer hand skills to the gripper of the root. Both the two robot arms work under torque control mode. The joints stiffness of the follower arm is set to be small ($e.g.$, $S_0 : 50, S_1 : 50, e_0 : 50, e_1 : 40, W_0 : 40, W_1 : 40, W_2 : 20$, N.m/rad), so the follower arm is soft when it is in physical contact with external objects. There is a large tracking error between the two arms, as shown in Fig. 3.32 (Right Top), which means that it has taken a very long time to complete the plugging-in task, as the human tutor has to compensate for the large tracking error via visual feedback. The contact force is comparatively smaller, as shown in Fig. 3.32 (Right Top). It indicates that the follower arm is capable of imitating human adaptation skills with a low contact force but with a large position error. Moreover, the robot failed to replay the task, which demonstrates the disadvantage of the fixed low gain stiffness for skill transfer.

In rigid mode, the process of the task execution is similar to that in soft mode. But the joints stiffness is set to be higher ($e.g.$, $S_0 : 400, S_1 : 400, e_0 : 300, e_1 : 300, W_0 : 200, W_1 : 200, W_2 : 50$, N.m/rad). The tracking error between the two arms is much smaller than the soft mode, as shown in Fig. 3.32 (Left Down). Therefore, it is easier to do the task without much position error compensation. However, as the follower arm is rigid, it will generate the high contact force, as indicated in Fig. 3.32 (Left Down), which may increase the possibility of changing the object's position or causing deformation. It indicates that the rigid mode with fixed high gain stiffness will lead to a smaller position tracking error but with much higher contact force, which is apparently not a good option for skill transfer in pHRI systems.

Plugging-in Task in the sEMG-based mode: In this mode, the plugging-in task is performed based on impedance control. The process is similar to the soft and rigid modes but with variable stiffness estimated based on the sEMG signals, which are extracted from the tutor's arm in the presence of haptic feedback in an almost real-time manner. Both stiffness profile and position trajectory are achieved for a complete skill transfer process. Fig. 3.32 shows the human tutor's stiffness adaptation, positions and forces of both the leader arm and the follower arm. It indicates that the tracking precision is much higher than that in soft mode, and the contact force of the follower arm is lower than that in the rigid mode, especially in the process of plugging into the socket. Due to the haptic feedback, the human tutor is able to adapt to the variation of the task by subconsciously adjusting his muscle activation, in order to obtain the desired stiffness. Therefore, the follower arm follows the movement of the leader arm with a better result (see the contact forces in the four modes). It means that a more stable and natural performance of skill transfer for the plugging-in task has been achieved in the developed sEMG-based mode.

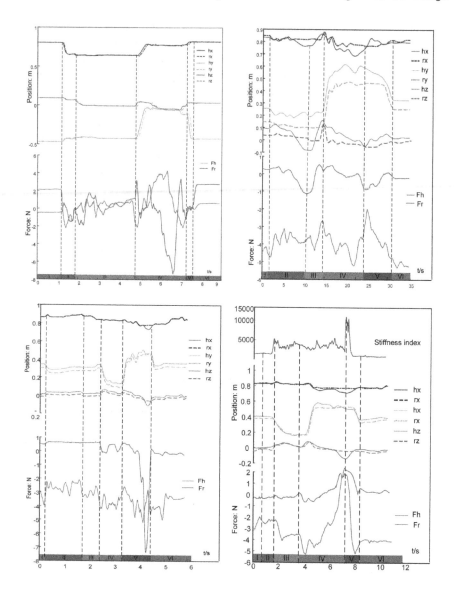

FIGURE 3.32 The results of the plugging-in task in default mode (Left Top), soft mode(Right Top), rigid mode (Left Down), and the sEMG-based mode (Right Down), respectively. The hx, hy, hz denote the positions of human tutor endpoint in x, y, z axis, respectively. The rx, ry, rz denote the positions of the robot endpoint in x, y, z axis, respectively. The Fh and Fr denote the tutor's endpoint force and the robot follower arm's endpoint force, respectively.

3.8 CONCLUSION

In this chapter, we have introduced several techniques for human-robot stiffness transfer based on sEMG signals. In these techniques, demonstrator's or tutor's variable stiffness is estimated through methods such as incremental stiffness estimation in 3D Cartesian space, the FSE and dimensionality reduction algorithms and one method using a combination of IMU and sEMG signals, and then transferred to the follower robot arm. Furthermore, techniques such as haptic feedback are used to aid. Experiments including several different tasks are carried out on a Baxter robot, and they have good performances with low position error and contact force, which may have potentials in human-robot coupling system such as telerehabilitation and exoskeleton.

REFERENCES

1. Ning Wang, Chenguang Yang, Michael R Lyu, and Zhijun Li. An EMG enhanced impedance and force control framework for telerobot operation in space. In *Aerospace Conference, 2014 IEEE*, pages 1–10. IEEE, 2014.
2. Arash Ajoudani, Nikos Tsagarakis, and Antonio Bicchi. Tele-impedance: Teleoperation with impedance regulation using a body–machine interface. *The International Journal of Robotics Research*, 31(13):1642–1656, 2012.
3. Yanbin Xu, Chenguang Yang, Peidong Liang, Lijun Zhao, and Zhijun Li. Development of a hybrid motion capture method using MYO armband with application to teleoperation. In *2016 IEEE International Conference on Mechatronics and Automation*, pages 1179–1184. IEEE, 2016.
4. Etienne Burdet, Rieko Osu, David W Franklin, Theodore E Milner, and Mitsuo Kawato. The central nervous system stabilizes unstable dynamics by learning optimal impedance. *Nature*, 414(6862):446–449, 2001.
5. A Ahmad Puzi, SN Sidek, and F Sado. Mechanical impedance modeling of human arm: A survey. In *IOP Conference Series: Materials Science and Engineering*, volume 184, page 012041, 2017.
6. Etienne Burdet, Gowrishankar Ganesh, Chenguang Yang, and Alin Albu-Schäffer. Interaction force, impedance and trajectory adaptation: by humans, for robots. In *Experimental Robotics*, pages 331–345. Springer, 2014.
7. Chenguang Yang, Gowrishankar Ganesh, Sami Haddadin, Sven Parusel, Alin Albu-Schaeffer, and Etienne Burdet. Human-like adaptation of force and impedance in stable and unstable interactions. *IEEE transactions on robotics*, 27(5):918–930, 2011.
8. Chenguang Yang, Zhijun Li, and Etienne Burdet. Human like learning algorithm for simultaneous force control and haptic identification. In *2013 IEEE/RSJ International Conference on Intelligent Robots and Systems*, pages 710–715. IEEE, 2013.
9. Yanan Li, Gowrishankar Ganesh, Nathanaël Jarrassé, Sami Haddadin, Alin Albu-Schaeffer, and Etienne Burdet. Force, impedance, and trajectory learning for contact tooling and haptic identification. *IEEE Transactions on Robotics*, 34(5):1170–1182, 2018.
10. Weibo Song, Xianjiu Guo, Fengjiao Jiang, Song Yang, Guoxing Jiang, and Yunfeng Shi. Teleoperation humanoid robot control system based on Kinect sensor. In *2012 4th international conference on intelligent human-machine systems and cybernetics*, volume 2, pages 264–267. IEEE, 2012.

11. Christopher Stanton, Anton Bogdanovych, and Edward Ratanasena. Teleoperation of a humanoid robot using full-body motion capture, example movements, and machine learning. In *Proc. Australasian Conference on Robotics and Automation*, 2012.

12. Maged S Al-Quraishi, Asnor J Ishak, Siti A Ahmad, Mohd K Hasan, Muhammad Al-Qurishi, Hossein Ghapanchizadeh, and Atif Alamri. Classification of ankle joint movements based on surface electromyography signals for rehabilitation robot applications. *Medical & biological engineering & computing*, 55(5):747–758, 2017.

13. Monica Tiboni, Alberto Borboni, Rodolfo Faglia, and Nicola Pellegrini. Robotics rehabilitation of the elbow based on surface electromyography signals. *Advances in Mechanical Engineering*, 10(2):1687814018754590, 2018.

14. Rieko Osu, David W Franklin, Hiroko Kato, Hiroaki Gomi, Kazuhisa Domen, Toshinori Yoshioka, and Mitsuo Kawato. Short-and long-term changes in joint co-contraction associated with motor learning as revealed from surface EMG. *Journal of neurophysiology*, 88(2):991–1004, 2002.

15. GC Ray and SK Guha. Relationship between the surface EMG and muscular force. *Medical and Biological Engineering and Computing*, 21(5):579–586, 1983.

16. Jörn Vogel, Claudio Castellini, and Patrick van der Smagt. EMG-based teleoperation and manipulation with the DLR LWR-III. In *2011 IEEE/RSJ International Conference on Intelligent Robots and Systems*, pages 672–678. IEEE, 2011.

17. Kyung-Jin You, Ki-Won Rhee, and Hyun-Chool Shin. Finger motion decoding using EMG signals corresponding various arm postures. *Experimental neurobiology*, 19(1):54–61, 2010.

18. Ji Won Yoo, Dong Ryul Lee, Yon Ju Sim, Joshua H You, and Cheol J Kim. Effects of innovative virtual reality game and EMG biofeedback on neuromotor control in cerebral palsy. *Bio-medical materials and engineering*, 24(6):3613–3618, 2014.

19. G Fabian Volk, M Finkensieper, and O Guntinas-Lichius. EMG biofeedback training at home for patient with chronic facial palsy and defective healing. *Laryngo-Rhino-Otologie*, 93(1):15–24, 2013.

20. Arash Ajoudani, Nikolaos G Tsagarakis, and Antonio Bicchi. Tele-impedance: Preliminary results on measuring and replicating human arm impedance in tele operated robots. In *2011 IEEE international conference on robotics and biomimetics*, pages 216–222. IEEE, 2011.

21. Atau Tanaka and R Benjamin Knapp. Multimodal interaction in music using the electromyogram and relative position sensing. *NIME 2002*, 2002.

22. Arash Ajoudani, Nikos Tsagarakis, and Antonio Bicchi. Tele-impedance: Teleoperation with impedance regulation using a body–machine interface. *The International Journal of Robotics Research*, 31(13):1642–1656, 2012.

23. Chenguang Yang, Chao Zeng, Peidong Liang, Zhijun Li, Ruifeng Li, and Chun-Yi Su. Interface design of a physical human–robot interaction system for human impedance adaptive skill transfer. *IEEE Transactions on Automation Science and Engineering*, 15(1):329–340, 2017.

24. J-U Chu, Inhyuk Moon, and M-S Mun. A real-time EMG pattern recognition system based on linear-nonlinear feature projection for a multifunction myoelectric hand. *IEEE Transactions on biomedical engineering*, 53(11):2232–2239, 2006.

25. Mahdi Khezri and Mehran Jahed. Real-time intelligent pattern recognition algorithm for surface EMG signals. *Biomedical engineering online*, 6(1):45, 2007.

26. Peidong Liang, Chenguang Yang, Ning Wang, Zhijun Li, Ruifeng Li, and Etienne Burdet. Implementation and test of human-operated and human-like adaptive impedance controls on Baxter robot. In *conference towards autonomous robotic systems*, pages 109–119. Springer, 2014.

27. Ryan J Smith, Francesco Tenore, David Huberdeau, Ralph Etienne-Cummings, and Nitish V Thakor. Continuous decoding of finger position from surface EMG signals for the control of powered prostheses. In *2008 30th Annual International Conference of the IEEE Engineering in Medicine and Biology Society*, pages 197–200. IEEE, 2008.

28. Jimson G Ngeo, Tomoya Tamei, and Tomohiro Shibata. Continuous and simultaneous estimation of finger kinematics using inputs from an EMG-to-muscle activation model. *Journal of neuroengineering and rehabilitation*, 11(1):122, 2014.

29. Cheng Fang and Xilun Ding. A set of basic movement primitives for anthropomorphic arms. In *2013 IEEE International Conference on Mechatronics and Automation*, pages 639–644. IEEE, 2013.

30. Chenguang Yang, Peidong Liang, Arash Ajoudani, Zhijun Li, and Antonio Bicchi. Development of a robotic teaching interface for human to human skill transfer. In *2016 IEEE/RSJ International Conference on Intelligent Robots and Systems (IROS)*, pages 710–716. IEEE, 2016.

31. David J Bennett. Stretch reflex responses in the human elbow joint during a voluntary movement. *The Journal of physiology*, 474(2):339–351, 1994.

32. Robert E Kearney, Richard B Stein, and Luckshman Parameswaran. Identification of intrinsic and reflex contributions to human ankle stiffness dynamics. *IEEE transactions on biomedical engineering*, 44(6):493–504, 1997.

33. Arash Ajoudani, Cheng Fang, Nikos G Tsagarakis, and Antonio Bicchi. A reduced-complexity description of arm endpoint stiffness with applications to teleimpedance control. In *2015 IEEE/RSJ International Conference on Intelligent Robots and Systems (IROS)*, pages 1017–1023. IEEE, 2015.

34. Andrea Maria Zanchettin, Nicola Maria Ceriani, Paolo Rocco, Hao Ding, and Björn Matthias. Safety in human-robot collaborative manufacturing environments: Metrics and control. *IEEE Transactions on Automation Science and Engineering*, 13(2):882–893, 2015.

35. John D Wason and John T Wen. Robot raconteur: A communication architecture and library for robotic and automation systems. In *2011 IEEE International Conference on Automation Science and Engineering*, pages 761–766. IEEE, 2011.

4 Learning and Generalization of Variable Impedance Skills

4.1 INTRODUCTION

It shows that stiffness adaptation plays a great role in human performing tasks [7]. Under the control of the Central Nervous System (CNS), humans are able to adapt the impedance of their biomechanical systems to different requirements of tasks and to external disturbances, to ensure robust and safe physical interactions with the environments [5, 8, 7]. Through muscle co-contraction, humans can naturally adapt to a complex task situation. This kind of task-specific impedance adaptation makes it possible for humans to combine the advantages of high stiffness and low stiffness control.

Similarly, the variable impedance regulation strategy is also necessary for a robot manipulator to acquire human-like skills [3, 9, 8, 10]. The application of variable impedance control has become one of the hot research topics in the robotics community. Similar to the natural co-contraction of human limb muscles, the obtained stiffness of a robot tends to be stable and continuously-adjustable. Thus, it is able to allow a robot to adapt to more complex task situations. This method has been successfully applied to a variety of task scenarios, such as plug-in-a-hole, ball-catching [4], wood-sawing [11], and drilling [12].

The ways to achieve variable stiffness can be roughly categorized into non-biological method [13] and sEMG-based method [4]. We believe that the later one has more advantages, one of which is that the stiffness profiles are directly obtained by monitoring and extracting humans limb sEMG signals. sEMG-based variable impedance transfer enables the robot to adapt the stiffness profiles to different task situations with a natural regulation process [1, 6].

From the perspective of programming efficiency, the robot is expected to generalize the learned skills to new given task situations. To this end, the trajectories (or behaviors) can be encoded by several models such as Dynamical Movement Primitives (DMP) [14], which has been widely applied to a number of tasks such as ball-in-a-cup [15], table tennis [16], and grasping [17]. Significantly, DMP is also used for encoding force profiles obtained from demonstrations in Refs. [13, 18, 19, 20]. In this chapter, we use DMP to encode both movement trajectories and stiffness profiles to achieve variable impedance skill transfer and generalization.

This chapter introduces a framework that enables robots to efficiently learn both motion and stiffness control policies from humans, in which human limb muscle activities are monitored for variable stiffness estimation, and the segmentation of both movement trajectories and the sEMG-based stiffness profiles are simultaneously realized based on the Beta Process Autoregressive Hidden Markov Model (BP-AR-HMM). Additionally, a dual-arm control strategy with haptic feedback mechanism is developed for skill transfer. Finally, experimental studies are conducted to verify the approach on a Baxter robot.

4.2 OVERVIEW OF THE FRAMEWORK

The proposed framework includes four phases: (i) demonstration, (ii) segmentation, (iii) alignment, and (iv) generalization. The overview of the framework is shown in Fig. 4.1. *Demonstration*: Conventionally, a human tutor demonstrates the skills to

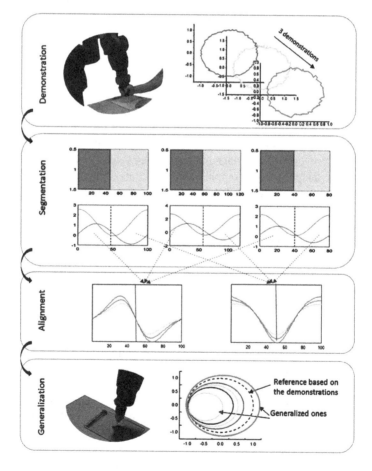

FIGURE 4.1   Graphical representation of the overview of the proposed framework.

accomplish one specific task, during which the movement trajectories are recorded for subsequent usage. In this framework, the demonstrator's arm sEMG signals are also extracted for the stiffness features. In order to capture as many features as possible, several times of demonstrations are often performed for one task. The dual-arm control-based teleoperation method described in Ref. [6] is used for demonstration. *Segmentation*: The skills represented by movement trajectories together with stiffness profiles obtained from the demonstrations are divided into sequences of subskills. In this way, a repository of features of one specific task is established. *Alignment*: The demonstration profiles with different time durations often need to be temporally aligned. Additionally, we also need to align the coordinate points from different segments. To this end, the Generalized Time Warping (GTW) algorithm [23] is utilized in this work.

4.3 TRAJECTORY SEGMENTATION

4.3.1 DATA SEGMENTATION USING DIFFERENCE METHOD

Difference Method (DM) is a numerical method for differential equations; it approximates the derivative by finite difference and seeks the approximate solution of the differential equation [21]. It is an approximate numerical solution of differential equations. Specifically, the difference method is to replace the differential with finite difference and to replace the derivative with a finite difference quotient, so that the basic equation and the boundary condition (generally the differential equation) are approximately changed to the difference equation (algebraic equation) [22]. The problem of solving differential equations is changed to solve the problem of algebraic equations.

One method for segmentation is to use DM. Considering the variable y_i depends on the independent variable z_i. When z_i changes to $z_i + 1$, the amount of change of the dependent variable $y_i = f(z_i)$, $df(z_i) = f(z_i + 1) - f(z_i)$ is called the difference of the function $f(z_i)$ with a step length of 1 at point z_i, often referred to as the difference of the function $f(z_i)$, and called d as the difference operator. The difference has an arithmetic property similar to the differential. The equation is shown as follows [21]:

$$f'(z_i) = df(z_i) = y_i' \approx \frac{y_{i+1} - y_i}{z_{i+1} - z_i} \tag{4.1}$$

where one significant factor of writing is that when each single stroke is completed, the pen would be picked up for once. Hence the "z" coordinate values from the experimental data are treated as the reference of the segmentation. $f(z_i)$ is the set of z_i values. After the DM, we have:

$$\xi(y_i') = sign(|y_i'| - \theta) \tag{4.2}$$

where ξ is the gaping factor; θ is a constant, and by giving different values of θ we can adapt the segmentation characters, such as the size of the segmented data set, here $\theta = 0.5$; *sign* is the Signum function, for each element of ξ, the formulation can be defined as follows:

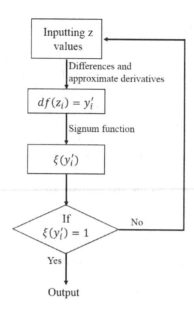

FIGURE 4.2   The flowchart of the segmentation.

$$sign(\xi) = \begin{cases} -1 & if \ \xi < 0, \\ 0 & if \ \xi = 0, \\ 1 & if \ \xi > 0. \end{cases} \qquad (4.3)$$

The $sign(\xi)$ values, which are equal to 1, correspond to those "z" coordinate data increasing sharply, where every single stroke writing ends. By collecting the values of $sign(\xi)$, we obtain several local text files for the usage of GMM and DMP generalization. Fig. 4.2 demonstrates the flowchart of the segmentation.

4.3.2   BETA PROCESS AUTOREGRESSIVE HIDDEN MARKOV MODEL

Hidden Markov Model (HMM) is a statistical method that is appropriate for time series analysis. It has been applied for encoding trajectory in robot skill acquisition. An HMM is a dual stochastic process that is characterized by a Markov chain of a sequence of hidden state variables and a corresponding sequence of observation data. Generally, a transition function is defined to describe the probability of each state at time t given the last state. Based on a set of observed data, the parameters of an HMM can be efficiently estimated by using the forward-backward or Viterbi algorithms, and the most likely sequence of states which generated the observed data can also be determined. However, a *priori* knowledge is often needed to choose an implicit number of states, easily leading to the problem of over-fitting or low-fitting. This drawback largely limits the utilization of the model, especially when dealing with movement segmentation of a complex task. Additionally, the conventional HMM is not well suitable to address the problem of segmenting complex behaviors according to multiple observed data because the observations are independently considered in the model, and the dependency between different series of observations are neglected [8].

The Beta Process Autoregressive HMM (BP-AR-HMM) can address these two

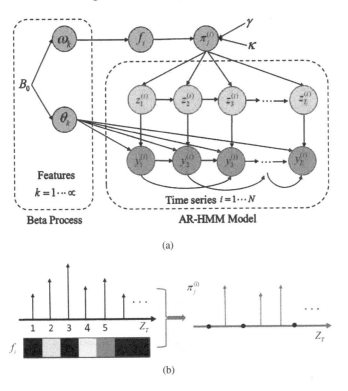

(a)

(b)

FIGURE 4.3 Graphical representation of the BP-AR-HMM model (a) and the feature selection mechanism (b).

drawbacks with the HMM model, as mentioned above. A graphical representation is shown in Fig. 4.3(a). It mainly contains two parts, i.e., the Beta process and the AR-HMM model. Here we briefly introduce the basic idea of the algorithm.

Firstly, B, generated by a Beta Process (BP), defines the global weights for the potentially infinite number of states, which can correspondingly encode a number of different behaviors, see equations (4.4) and (4.5). The feature-inclusion probabilities ω_k and state-specific parameters $\theta_k = \{A_k, \Sigma_k\}$ are generated via this process [2]:

$$B|B_0 \sim BP(1, B_0) \qquad (4.4)$$

$$B = \sum_{k=1}^{\infty} \omega_k \theta_k \qquad (4.5)$$

Then, a Bernoulli Process (BeP) parameterized by B is performed to generate an X_i for each time series i, as shown in equation (4.6). Each X_i is utilized to generate a binary vector $f_i = [f_{i1}, f_{i2}, ...]$ indicating which of the global features are shared in the ith time series. For instance, if the kth element in the vector f_i is 1, i.e., $f_{ik} = 1$, then the ith time series includes the jth feature. Therefore, this enables the possibility of sharing of global features amongst multiple time series; meanwhile, it allows variability for each time series. The graphical representation of the feature selection mechanism is shown in Fig. 4.3(b).

$$X_i | B \sim \mathrm{BeP}(B) \tag{4.6}$$

$$X_i = \sum_{k=1} f_i \delta_{\theta_k} \tag{4.7}$$

Thirdly, the transition distributions $\pi_j^{(i)}$ is constructed via a Dirichlet distribution with the hyper-parameters γ and κ, given the feature-indicating vector f_i for each time-series. Thus, the ith time-series can select features from a global library with the feature-constrained transition distributions $\pi_j^{(i)}$.

$$\pi_j^{(i)} | f_i, \gamma, \kappa \sim \mathrm{Dir}([\gamma, ..., \gamma + \kappa, \gamma, ...] \otimes f_i) \tag{4.8}$$

Fourthly, the state $z_t^{(i)}$ is generated for each time step t based on the transition distribution of this mode at the last time step, as shown in equation (4.9).

$$z_t^{(i)} \sim \pi_{z_{t-1}^{(i)}}^{(i)} \tag{4.9}$$

Finally, the observations $y_t^{(i)}$ can be obtained using equation (4.10), which shows a Vector Autoregression Model (VAR). It demonstrates that given the VAR order r, the observation is calculated as the sum of linear transformations of the previous r observations of the model, plus a state-specific noise.

$$y_t^{(i)} = \sum_{j=1}^{r} A_{j, z_t^{(i)}} y_{t-j}^{(i)} + e_t^{(i)}(z_t^{(i)}) \tag{4.10}$$

The BP-AR-HMM model is able to provide reliable inference with only a few open parameters. A Beta process prior is used to determine the total number of the potential features representing a complex task in a fully Bayesian way, without the need for manual intervention, as mentioned above. More importantly, it is able to recognize global features among multiple demonstrations, as well as to leave room for the variability of each individual time series.

In this work, we will use the model to segment not only the movement trajectories obtained during multiple demonstrations but also the estimated human arm stiffness profiles based on sEMG signals obtained from the human demonstrator. To this end, we encapsulate the movement information x and the stiffness information p into the observations: $y_t^{(i)} = \{x_t^{(i)}, p_t^{(i)}\}$ such that they can be segmented in parallel.

4.4   TRAJECTORY ALIGNMENT METHODS

To realize the multi-modal alignment of human motion, the work [23] proposed the GTW algorithm, which can address the issues above. Compared to the DTW, GTW has three advantages: (i) it uses a Gauss-Newton algorithm with linear complexity in the length of the sequence to optimize the time warping function; (ii) it considers the differences in dimensionality using multi-set canonical correlation analysis (mCCA); (iii) it uses a more flexible warping model parametrized by a set of monotonic bases to compensate for changes in time space.

Considering m demonstration profiles, $\{U_1, \cdots, U_m\}$, with $U_i = [u_1^i, \cdots, u_{n_i}^i]$, GTW minimizes the cost function [23]:

$$J_{gtw} = \sum_{i=1}^{m} \sum_{j=1}^{m} \frac{1}{2} \|V_i^T U_i W_i - V_j^T U_j W_j\|^2 + (\sum_{i=1}^{m} \varphi(W_i) + \phi(V_i)) \qquad (4.11)$$

where W_i and V_i are the non-linear temporal transformation and the low-dimensional spatial embedding, respectively. $\varphi(\cdot)$ and $\phi(\cdot)$ are regularization functions. Equation (4.11) suggests that for each U_i, GTW finds a W_i and a V_i such that the sequence $V_i^T U_i W_i$ is well aligned with the others in the least-squares sense. This is a non-convex optimization problem with respect to the alignment (i.e., W_i) and projection matrices (i.e., V_i), which are solved by Gauss-Newton algorithm and multi-set canonical correlation analysis, respectively.

4.5   DYNAMICAL MOVEMENT PRIMITIVES

Typically, a point-to-point movement DMP for a one-dimensional system is represented by three differential equations [14, 8]:

$$\tau \dot{v} = K(x_g - x) - Dv + (x_g - x_0)f(s; w) \qquad (4.12)$$

$$\tau \dot{x} = v \qquad (4.13)$$

$$\tau \dot{s} = -\alpha_1 s \qquad (4.14)$$

Here, we ignore the time variable, i.e., we denote x_t as x. Equation (4.12) represents a transform system which consists of a linear dynamical system acting like a spring-damping system perturbed by a non-linear forcing function, i.e., $f(s; w)$. K and D are constant stiffness coefficient and constant damping coefficient, respectively. And D is usually set to be $D = 1/4K$ such that the linear dynamical system is critically damped. x_g and x_0 are the goal and initial point of a trajectory, respectively. τ is a scaling factor shared by all these equations. x and v denote the position and velocity of the trajectory, respectively, and they are related, as shown in equation (4.13). Finally, s is the phase variable of the transform system, and it is represented by another differential equation, see equation (4.14), and the α_1 is a predefined positive constant.

The non-linear forcing function $f(s; w)$ takes the following representation:

$$f(s; w) = w^T g \qquad (4.15)$$

where g is a time-parameterized kernel vector, and w is the policy parameter vector, which affects the shape of the learned trajectory. The element of the kernel vector is defined as:

$$[g]_n = \frac{\varphi_n(s)s}{\Sigma_{n=1}^{N} \varphi_n(s)} \qquad (4.16)$$

with a normalized basis function $\varphi_n(s)$ which is usually defined as a radial basis function (RBF) kernel:

$$\varphi_n(s) = \exp(-h_n(s - c_n)) \qquad (4.17)$$

where c_n and h_n are the centers and widths of these kernel functions, respectively. Generally, c_n are equispaced in time on the whole trajectory, and h_n are selected based on experience.

The policy parameter w can be efficiently learned using supervised-learning algorithms such as weighted linear regression (LWR). To put it simply, we need to find a proper parameter vector by minimizing the following error:

$$\min \sum (f_{target} - f(s))^2 \qquad (4.18)$$

with

$$f_{target} = \frac{\tau \dot{v} + Dv - K(x_g - x)}{x_g - x} \qquad (4.19)$$

which can be computed based on a demonstrated trajectory $x_t, \dot{x}_t, \ddot{x}_t$, with $t = 1, \cdots, T$ and $x_g = x(T)$. To learn the weights from multiple demonstrations, we take a direct way to determine the weights. First, each demonstration is evaluated with a score of the scale from one to ten according to its corresponding performance. For each demonstration, a coefficient κ_i is related to the scores π_i and defined as follows

$$\kappa_i = \frac{\pi_i}{\Sigma_{i=1}^{L} \pi_i} \qquad (4.20)$$

where L is the total number of the demonstrations. Then, the final DMP output $\Omega = \{x, \dot{x}, p\}$, is computed by

$$\Omega = \sum_{i=1}^{L} \kappa_i \Omega_i \qquad (4.21)$$

Thus, the multiple demonstrations can be integrated together to yield a final output. Then, a direct way of computing the mixture of movement primitives (see Eqs. (4.20) and (4.21)) is used to generate one output of the DMP model based on the multiple demonstrations.

4.6 MODELING OF IMPEDANCE SKILLS

In the existing DMP model, a transform system, which is used to represent movement trajectory, is driven by a Canonical system. The output of the model only involves kinematics, i.e., positions, velocities, and accelerations. Thus, it is not possible to realize impedance control using the model. While we believe that movement and stiffness are equally important for a robot manipulator to acquire human-like skills, especially in physical human-robot interaction scenarios, and they should be equally accounted for Ref. [7]. One feasible and efficient way to achieve this is to add another transform system to represent the stiffness profile. The extended framework of the DMP model is shown in Fig. 4.4. The extended DMP framework for a one-dimensional state variable is represented as follows:

$$\dot{s} = h(s) \tag{4.22}$$

$$\dot{x} = g_1(x, s, \gamma) \tag{4.23}$$

$$\dot{p} = g_2(p, s, \omega) \tag{4.24}$$

where $h(s)$ denotes the Canonical system, which is defined as bellow:

$$\tau\dot{s}(t) = -\alpha s(t) \tag{4.25}$$

where α is a predefined constant. The s is the state of a first-order dynamic system, i.e., the canonical system. The state s can be seen as a phase variable which is set to monotonically decrease from 1 to 0. Eqs. (4.23) and (4.24) represent the transform

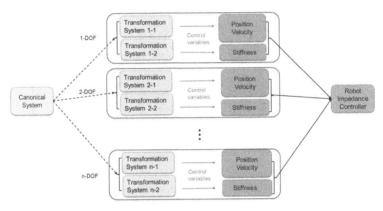

FIGURE 4.4 The extended framework of the DMP model. For each DOF, two sets of transform systems are used to represent for motion trajectories and stiffness profiles respectively. Reference movement trajectories and reference stiffness profiles are used to train the systems. Each DOF has its own nonlinear function and transformation systems, but they has the same canonical system. All the transform systems are driven by the same Canonical system using the same temporal scaling coefficient to grantee phase synchronization. The generated control variables are finally mapped into the robot impedance controller.

systems, which are used to encode movement trajectory and stiffness, respectively. $\{x, p\}$ denote the "positions" of movement and stiffness in time steps, and $\{\dot{x}, \dot{p}\}$ denote the "velocities" of that in time step accordingly. $\{\gamma, \omega\}$ are the internal parameters of the two transform systems, respectively. Note that these two transform systems are driven by the same Canonical system such that the phase synchronization can be guaranteed.

4.7 EXPERIMENTAL STUDY

4.7.1 LEARNING WRITING TASKS

A KUKA LBR iiwa robot is used in this experiment. The experimental procedure is as follows: the demonstrator teaches the robot to write the first half of a Chinese poem by holding the marker pen attached in the end-effector of the KUKA robot. Meanwhile, the trajectories data are outputted locally for the usage of programming. The trajectories data are then sent to the remote desktop computer to be segmented using MATLAB. The principle is that the demonstrator needs to pick up the marker pen after the writing of every single stroke. And the data outputted by the robot are a set of the coordinates of the end-effector, including the values of X, Y, Z, A, B, C. Wherein the A, B, C values are the Euler angles of the end-effector. Hence, we can find the various range of Z value, if it is suddenly increasing apparently, it means that a stroke has been finished. After the segmentation of experimental

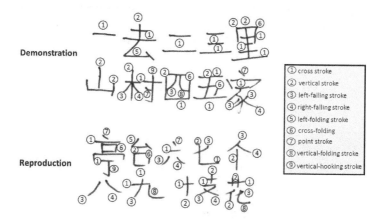

FIGURE 4.5 The demonstrated Chinese characters and the DMP generalization based play-back Chinese poem characters with their strokes contents.

data, the split groups of data are sent for optimization based on GMR. By applying GMR, all the repetitive strokes can be optimized into a single stroke. Although there are only ten Chinese characters in the first half of the poem, almost all the strokes are exported without any repetition after the application of GMR. Then, those data are encoded with the DMP model to generalize. By doing this, all the strokes can be adapted accordingly, which are relied on to re-group the new Chinese characters. As shown in Fig. 4.5, twenty separate Chinese characters are written by teaching, and wherein there are half characters written by teaching and half by the DMP generalization-based playback process. There are a total of nine strokes cross in the first half of the Chinese poem written by the KUKA robot under the teaching of a demonstrator, which are vertical stroke, left-falling stroke, right-falling stroke, cross-folding stroke, left-folding stroke, point stroke, vertical-folding stroke and vertical-hooking stroke, respectively. It can be concluded from Fig. 4.5, that the strokes contents of generated Chinese characters from the DMP-based generalization are the same as those that taught by the demonstrator, and that they correspond to each other one to one. To be more precise, all the strokes segmented from the recorded trajectories of the teaching process are able to re-group those characters written during the playback process of the KUKA robot after the application of GMR and DMP generalization. For example, the upper part of the second character "tai" in the second half of the poem is made by the lower part of the second character "qu" in the first line of the poem; the lower part of the second character "tai" in the second half of the poem is made by the outer part of the third character "si" in the second line of the poem, where the size and the position of the character parts are adjusted with the DMP model.

GMR is used to optimize all the repetitive strokes into a single one, such as the horizontal stroke, which is the most repetitive stroke. As a consequence, after some experiments, the GMM number is able to result in the influence for the forming of optimal trajectories. Furthermore, we choose the GMM number as 20, and it is obtained after a number of training, to achieve great performances for our experimental section. Through the employment of the GMR, the restoration of data during the TbD has been converted into a process, where we use GMR algorithm to estimate the joint distribution, which can be approximated by a mixture of Gaussian functions. During the calculation, the certain correlation of the points in the sample set of linear data is significant for the learning procedure of robots. On the basis above, the GMM number decides the prediction process. The generated trajectories (the last two lines) are then drawn by the KUKA LBR robot arm over the recorded patterns to provide a visual result (shown in Fig. 4.5), which can be easily compared with the pre-written Chinese characters (the first two lines).

4.7.2 PUSHING TASKS

Two experiments are used to evaluate the performance of the DMP while learning from multiple demonstrations [25]. These experiments are carried out on a Baxter robot in V-REP. In order to avoid the correspondence problem [26], we use computer vision to capture the human tutor's action and then transmit the demonstrations information to the robot. We select the Kinect as the motion-sensing input device. In our experiments, the robot learns the human demonstrations in joint space. Therefore, the states of the human tutor's joints are captured by the Kinect, and the shoulder joints and the elbow joint is used in the experiments. (see Fig. 4.6 Left).

The first experiment is to let the robot learn how to push the board on the left (for the robot) and reproduce the learned motion. As shown in Fig. 4.7, the Baxter robot imitates the human tutor's motion to perform the task. The robot raises its left arm over the pillar and then moves its gripper to push the left board off the table. This demonstration is repeated 10 times. All of the joint angles that we focus on are recorded throughout the demonstrations. And then the data is used to train the DMP. In order to match the autonomy of the dynamic system, we use time steps to represent the duration of the motion, and all durations of the demonstrations are converted to 100 steps. The learning result is shown in Fig. 4.8. For each joint angle, the trajectory

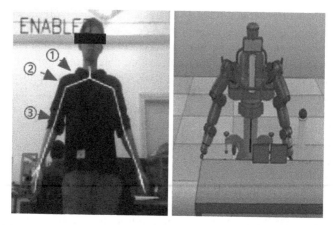

FIGURE 4.6 Left: the human tutor. Three joint states of the left arm are captured. (The figure captured by the Kinect is the mirror of the reality.) Right: the initial scene of the experiments.

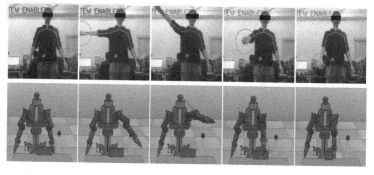

FIGURE 4.7 Top: the human tutor demonstrates how to push the left board off the table. Bottom: the robot imitates the motion of the human tutor.

of the reproduction is similar to the demonstrations in shape. We test the generated motion on the Baxter robot without modulating the start or the goal of the motion. As shown in Fig. 4.9, the robot pushes the left board off the table successfully without knocking down the pillar and the ball.

The original DMP is learned from a single demonstration. In order to compare the performance of two methods, we use those demonstrations to train 10 DMP. We use them to generate new motions without modulating the goal. Then we apply them on the robot to test if the robot is able to perform the task successfully. Six of them can not push the left board off the table or will push both boards. After modulating the goal of these DMP to an appropriate position, the robot can also complete the task with the regenerated motions.

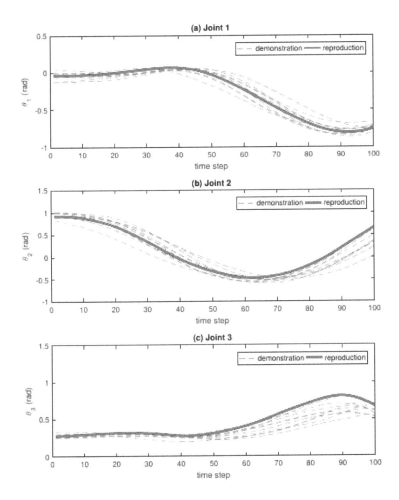

FIGURE 4.8 The demonstrations (dashed-line) and the reproductions (solid line) of three joint angles.

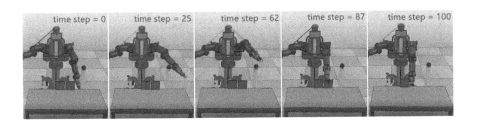

FIGURE 4.9 The motion generated from the learned DMP.

In the second experiment, we modulate the initial angle of the joint 1 to evaluate the stability of the DMP. As shown in Fig. 4.10, the modulated motions are still stable at the goal. To evaluate the spatial scaling ability of DMP, we modulate the goal of the motion to another board. As Fig. 4.11 shows, the previous goal of three joint angles is $[\theta_1, \theta_2, \theta_3] = [-0.755, 0.652, 0.664]$ (rad). We modulate it to $[-0.987, 0.564, 0.635]$ so that the left arm of the robot can reach the right board. The evolution of three joint angles is as shown in Fig. 4.11. The trajectories keep their shape and finally get to the goals accurately. We also applied the generated motion on the robot. As Fig. 4.12 shows, the robot's left arm moves around the pillar and then push the right board successfully. Another ability of DMP is temporal scaling. Real-time temporal-spatial factor τ, from 1 to 0.5, can speed up the generated motion. Three joint angles get to the goals at time step $= 50$ in Fig. 4.13. As shown in Fig. 4.14, the robot completes the task at twice the speed.

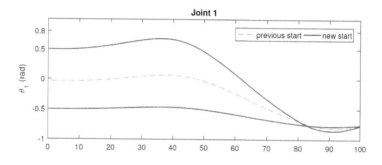

FIGURE 4.10   The evolution of the joint 1 angle while modulating the initial angle.

4.7.3   CUTTING AND LIFT-PLACE TASKS

A Baxter robot is used in our experimental study. The human tutor's muscle activity is monitored using a MYO device to collect raw sEMG signals. Two tasks, cutting and lift-place, are illustrated to justify the effectiveness of the proposed method [27]. As shown in Fig. 4.15, one arm of the robot is used as the leader arm, which is physically attached to the human tutor's limb endpoint through a specially designed coupling device [6, 27]. The other arm serves as a follower arm, and a knife is mounted onto its endpoint as a cutting tool. In the teaching phase, the human tutor drives the leader arm to move, and the follower arm follows the movement and learns the patterns of the position trajectory and the stiffness profile of the hand of the human tutor. While in the reproduction phase, the follower arm is able to reproduce and generalize the learned skill without the assistance of the leader arm.

First, in the teaching phase, the human tutor who has been involved in the stiffness calibration demonstrates the cutting skill through teleoperation ten times. The trajectories of the seven joints, as well as the tutor's limb sEMG signals, are recorded

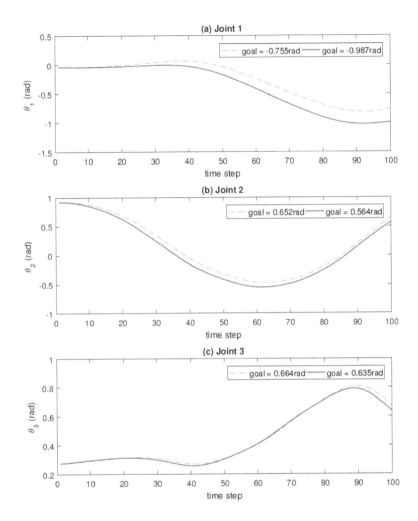

FIGURE 4.11 The evolution of three joint angles while modulating the goal position to another board.

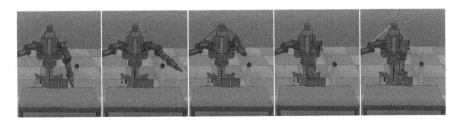

FIGURE 4.12 The motion generated while modulating the goal position to another board.

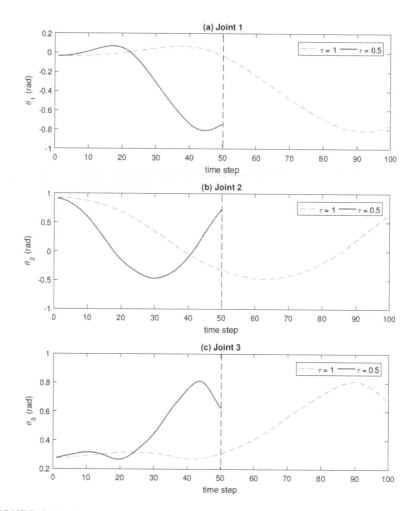

FIGURE 4.13   The evolution of three joint angles while modulating the temporal scaling factor τ from 1 to 0.5.

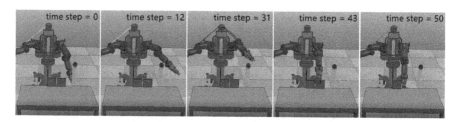

FIGURE 4.14   The motion generated while modulating the temporal scaling factor τ from 1 to 0.5.

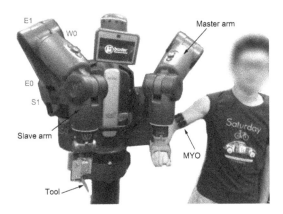

FIGURE 4.15 The experimental setup for the cutting task during demonstration.

during each demonstration. Note that the sEMG signals can be evaluated in an almost real-time manner. Thus we can directly record the sEMG data and calculate the stiffness indicator p.

A qualitative criterion is employed to determine the weight of every single demonstration. To that end, each one of three subjects gives a score ranging from one to ten for each signal demonstration according to the task execution performance. If the system is unstable when the follower arm contacts with the object or the object is not cut well eventually during the demonstration, it will get a lower score, the weights are calculated by using Eq. (4.20). One may also employ objective approaches for learning from multiple demonstrations (see, e.g., Ref. [28]).

The spring coefficient k and damping coefficient d of the DMP model is set to 150 and 25, respectively. The numbers of basis functions are chosen: 100 and 50 for the trajectories DMP and the stiffness DMP, respectively. Once we achieve the generated movement trajectories and stiffness profiles that account for multiple demonstrations, the following two subtasks are conducted:

In this subtask, we would like to extend the ability of DMP model for skill generalization spatially from position control mode to impedance control mode. The goal is to generate the learned skill to cut an object in different spatial positions. In our case, it is designed to cut a cucumber in different positions along the y-axis. The robot here is controlled in sEMG-based variable impedance mode. Note that in position control mode, only paths need to be planned, while in impedance control mode, both the paths and velocities should be adjusted. The generalized trajectories and velocities are achieved just by changing the profile of the DMP model, and they are shown in the *top row* in Fig. 4.16. It demonstrates that the generalized trajectories are very similar to the reference ones but converging to different goals.

The results of this subtask are shown in the *bottom row* in Fig. 4.16. Fig 4.16 (c) shows the measured endpoint position in the y-axis of the robot follower arm. It

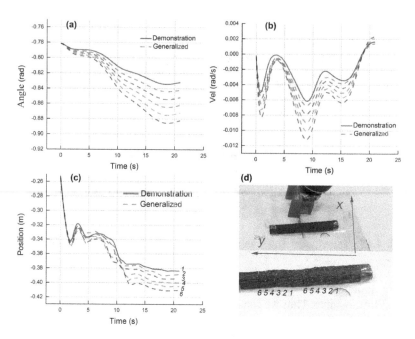

FIGURE 4.16 *Top*: the planned trajectories for skill spatial generalization. *Bottom*: (c) is the measured robot endpoint position in *y* axis; (d) shows the executed cut locations on the cucumber, two sets of numbers (1–6) mean this subtask are successfully performed twice (six cuts once). Cut No.1 refers to the demonstrated one, and the others refer to the generalized ones. The solid lines denote the reference trajectories obtained from imitation learning. And the dash lines denote the generalized ones.

demonstrates that the endpoint could generalize spatially to different positions. Fig 4.16(d) shows the executed cuts. This subtask demonstrates that DMP model for skill spatial-generalization can be successfully realized in the impedance control model.

One of the most important contributions of our proposed framework lies in the ability to generalize the stiffness profiles to new given situations. This subtask aims to demonstrate this point. To that end, we generalize the reference stiffness indicator (the reference terminal value is 0.65) to two different goals, e.g., 0.8 and 0.9, based on the extended DMP model. It is shown in Fig. 4.17(a). A different object is used for cutting in this subtask, which has a thicker skin and larger hardness than the object used in the teaching phase and the subtask above.

The robot is required to perform the subtask five times under each one of the three conditions described above, resulting in a success rate of 0/5, 1/5, and 5/5, respectively. Fig. 4.17(b–f) show one typical result of this subtask. In this case, it shows that the object cannot be successfully cut using the impedance control mode with the reference variable stiffness. And when the final stiffness value is generalized to 0.8, the subtask still cannot be successfully finished. Until it is generalized to a higher value, i.e., 0.9, the object can be successfully cut. Note that there is a gap

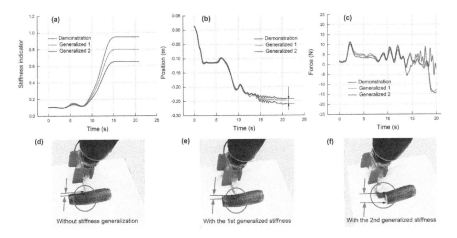

FIGURE 4.17 *Top* : (a) is the generalization of the stiffness indicator profile to different goals. (b) and (c) are the measured positions and forces of the follower arm end-effector in z-axis, respectively. *Bottom* shows the different cuts by using the reference (d), the first generalized (e), and the second generalized stiffness indicator profiles (f), respectively. The object can be cut successfully only by applying the second generalized stiffness profile. Note that the object used to cut here is different from the object used in the first subtask as shown in Fig. 4.16.

between the final positions of the reference and the second generalized trajectories, which represents the height of the object (Fig. 4.17(b)). From the perspective of force (Fig. 4.17(c)), it can be seen that the robot lacks adaptability even in variable impedance control mode if the desirable stiffness profile is not properly generalized, which is consistent with our common experience when people perform tasks.

The setup for the lift-place task is shown in Fig. 4.18. There are two different objects used in this task, object 1 and object 2 have the same size (10 cm × 10 cm × 10 cm), but different weights: 0.83 kg and 1.40 kg, respectively. The experimental procedure of this task is similar to the cutting task. During the demonstration, the same human tutor teaches the robot to lift the object 1, pass over the obstacle (with a height of 18 cm), and finally place it on the goal (the red cross mark). The human tutor is required to lift the object at as low height as possible to pass it over the obstacle for energy saving. After a given number of demonstrations, we can generate one integrated position trajectory and one integrated stiffness profile, as shown in Fig. 4.19. The parameters of the DMP model are set as the same as in the cutting task.

During the playback phase, the robot is required to perform the task under three different situations: Subtask 1: to lift and place the object 1 to a new given goal (the black cross mark), which is 5 cm away from the old one, with the demonstrated stiffness profile (i.e., trajectory spatial-schedule); Subtask 2: to lift and place the object 2 to the new given goal with the demonstrated stiffness profile (i.e., without stiffness schedule); and Subtask 3: to lift and place the object 2 to the new goal with a generalized stiffness profile (i.e., both trajectory and stiffness generalization).

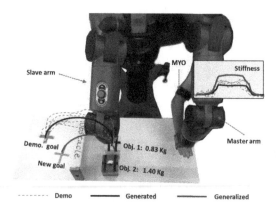

FIGURE 4.18   The experimental setup for the lift-place task.

Each subtask has been performed seven times. A success rate of $7/7$ is obtained in the first subtask, which shows that the spatial skill generalization has been successfully realized. The robot has finished the second task and the third subtask with a success rate of $1/7$ and $6/7$, respectively. The two comparative subtasks demonstrate the effectiveness of the proposed framework for stiffness schedule. Fig. 4.20 I $\sim$ IV show one typical example of the demonstrations and the three subtasks, respectively.

It should be emphasized that subtask 2 in III-(a) (as shown in Fig. 4.20) and subtask 3 in III-(b) are successfully performed just by adapting the parameters of the DMP model instead of teaching the robot again or requiring additional time-consuming process (e.g., [29], [30]) for achieving the proper stiffness profiles. Additionally, different combinations of the trajectories and the stiffness profiles can meet different requirements of task situations based on the proposed framework. More specifically, by combining different movement trajectories with the same stiffness profile, skill spatial generalization under variable impedance control mode can be realized (subtask 1 in III-(a) and subtask 1 in III-(b)), and by combining different stiffness profiles with the same movement trajectory, then stiffness schedule can be realized (subtask 2 in III-(a)), or both trajectories and stiffness profiles can be simultaneously achieved (subtask 3 in III-(b)). Furthermore, the DMP-based framework is model-free and available for different robotic platforms.

The reference stiffness profile for stiffness generalization is obtained based on the extraction of the sEMG signals and the subsequent calculation for the endpoint stiffness estimates. The human endpoint stiffness estimation method used in this work has been proved useful for impedance regulation in Ref. [4]. Therefore, on the one hand, the stiffness transfer and generalization are achieved conveniently and efficiently; on the other hand, the reference stiffness profile largely depends on the accuracies of the extracted sEMG signals and the stiffness estimation model. One may

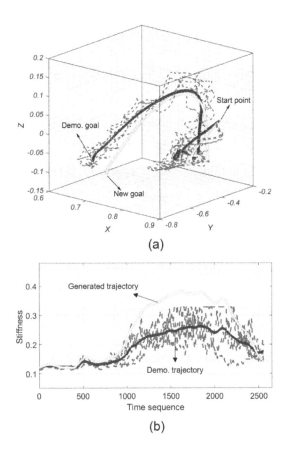

FIGURE 4.19   The trajectories (a) and stiffness profiles (b) of the lift-place task.

employ a much more complete stiffness estimation model to improve the accuracy of the estimated reference stiffness profile since the calculation efficiency of the estimation during the demonstration phase is not a strict constraint in this context.

Regarding the way, multiple demonstrations are used to generate a reference trajectory and a stiffness profile (see Fig. 4.19). The weight of every single demonstration is determined in a qualitative way for simplicity. Other more objective methods such as Probabilistic Movement Primitives (e.g., [31]) can be utilized to achieve the same purpose. Moreover, different human tutors have different demonstrations during the skill transfer process. Thus, it may help robots to achieve a better performance of executing tasks by considering the variance between different humans, namely learning from different human tutors.

The significance of stiffness adaptation for human-to-robot skill transfer has been verified in [6]. The adaptive stiffness process of the cutting task and the lift-place task can be seen in Figs. 4.17(a) and 4.19(b), respectively. This work introduces the

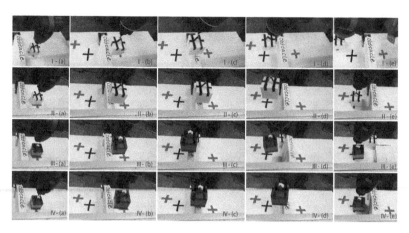

FIGURE 4.20 The experimental results of the lift-place task: I-(a)~(e), II-(a)~(e), III-(a)~(e), and IV-(a)~(e) show one typical example of the demonstrations, subtask 1, subtask 2, and subtask 3, respectively. The robot fails to perform subtask 2.

capability of stiffness generalization, therefore enriches the diversity of the human-to-robot variable impedance transfer. In our tasks, the stiffness generalization refers to the regulation of the stiffness profile according to the task requirement. For example, during the reaching step in the lift-place task, the stiffness is expected to remain small, it increases when larger stiffness is required to lift a heavier object. Surely, it can be decreased accordingly, as shown in Fig. 4.19(b).

4.7.4　WATER-LIFTING TASKS

The commercial sEMG armband named MYO is used to collect raw sEMG signals for monitoring the human demonstrator's muscle activities and stiffness estimation. The water-lifting task has been performed in this work [8]. The experimental setup is shown in Fig. 4.21.

During the demonstration, the leader arm of the Baxter robot is directly collected to one of the human demonstrator's arms through a physical coupling interface. The tutor demonstrates the water-pouring skill, and the robot follower arm follows the motion of the leader arm because of the virtual spring-damping system attached between these two arms, see Ref. [6] for details.

As shown in Fig. 4.22, the MYO worn on the demonstrator's upper arm is used for collecting the sEMG signals in a real-time manner. The sEMG signals are sent to a computer using Bluetooth transmission and then processed for the estimation of human arm endpoint stiffness. Subsequently, the estimated stiffness profile is properly mapped into the robot arm impedance controller using the UDP protocol to generate robot motion control commands. Meanwhile, the robot arm state and stiffness are recorded with a sample rate of 100 Hz.

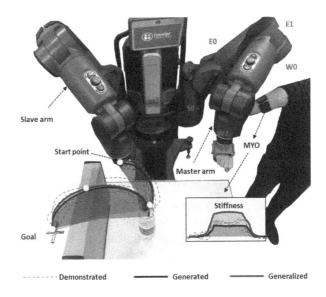

FIGURE 4.21   The experimental setup for the water-lifting task.

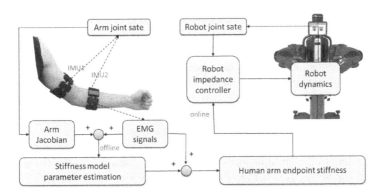

FIGURE 4.22   The mapping of human arm endpoint stiffness to robot joint impedance controller.

A BP-AR-HMM MATLAB implementation was first made available by E. Fox to segment sets of motion features collected from multiple demonstrations. Then, S. Niekum modified it to make it more suitable for the segmentation of movement trajectories in PbD. During the segmentation process, only the demonstrated movement trajectories and stiffness profiles in joint space are given without any other knowledge about the task. Following the instruction by Emily Fox, the obtained demonstration profiles are preprocessed such that the variance of the first differences in each dimension is 1, and the average value is 0. There are a total of 8 dimensions: 7-DoFs movement trajectories and 1-DoFs stiffness profile. Here, the variability of the stiffness is represented by a stiffness indicator, see Ref. [6]. All the demonstrated profiles are also subsampled down to 20 Hz and smoothed, as suggested in Ref. [32].

We choose the autoregressive order of 1 in AR-HMM model and adopt the other parameters the same as the parameters chosen by Emily Fox. Considering the principled statistical technique of the BP-AR-HMM model, the Metropolis-Hastings and Gibbs sampler is run 10 times for 1200 iterations each to segment the demonstration profiles, producing 10 segmentations. We chose the segmentation, which has the highest log-likelihood of the feature settings of the 10 runs and chose the segments as the movement primitives for the follow-up treatment.

In the demonstration phase, the human tutor demonstrates the water-lifting task six times. Unlike Refs. [32, 33], in this experiment the start point and the goal are both fixed in all demonstrations such that the movement trajectories are not largely different in shape, this is because we are only interested in considering the stiffness variability. To this end, the robot is taught to perform the task under two conditions: (i) to lift *half bottle of water* four times, and (ii) to lift *one-third bottle of water* twice. It is understandable that the six demonstrations have different lengths in time space. Additionally, the stiffness profiles are not exactly the same as each other in shape, even under the same task condition due to the uncertainty of the human demonstrator's muscle activation. The result of the segmentation is shown in Fig. 4.23. The top row shows segmenting the demonstration profiles into sub-skills marked with different colored bars, and the bottom row shows the corresponding divisions overlaid on a subplot of each of the 8-DoFs demonstration profiles. The first, third, fifth and sixth subplots correspond to the first task condition, while the second and the fourth subplots correspond to the second task condition. It can be seen that the BP-AR-HMM can successfully segment the water-lifting skill into four sequences basically corresponding to the multiple steps of the task: (i) reach the bottle; (ii) pick up the handle of the bottle; (iii) get close to the obstacle; (iv) lift the bottle to pass over the obstacle and place it to the goal. The first and third phases are considered as the same sub-skill (colored with light green). This is mainly because these two steps of the task have similar characteristics, only differing in reaching to different objectives. The BP-AR-HMM is able to consistently recognize the repeated sub-skills across the multiple demonstrations, even though they occur at different positions and with different stiffness. The segments are formally identical with minor defects in

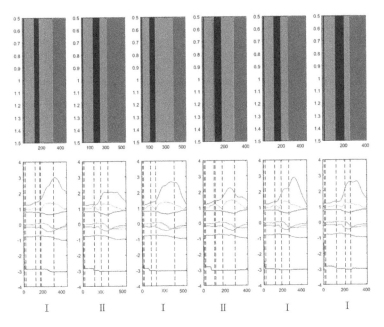

FIGURE 4.23 The result of segmentation of the 7-DoFs movement trajectories and the 1-DoF stiffness profiles. The first line in each of the six subplots denotes the stiffness, and the others represent the joint angles. I and II correspond to the first and the second task conditions, respectively.

the first and the fourth demonstrations which might be caused by the sudden changes of movement trajectories. These minor defects can be directly neglected since they do not affect much the performance of the segmentation.

Fig. 4.24 shows the result of the alignment of the demonstration profiles. The GTW can align each dimensional profiles all at once, thus resulting in a high alignment efficiency. The gray lines and the red lines denote the aligned profiles and the generated profiles, respectively. The aligned profiles are averaged and then modeled with the DMP model. One may also utilize some statistical approaches to encode the multiple profiles (see, e.g., Ref. [34]). It is seen that all the demonstration profiles, the 7-DoFs movement trajectories and the 1-DoF stiffness are well aligned in time space. The visual inspection of the lines in Fig. 4.24 suggests that the generated ones can capture the features in each dimension across the six demonstrations. Additionally, GTW is also able to align all lengths of the profiles to a particular value, which can facilitate the task plan.

Subsequently, we examine the task generalization to the following task situations: (i) to lift a bottle *full of water* to pass over the obstacle and place it to the goal; and (ii) *to keep low stiffness control as demonstrated* when high stiffness is undesired, namely, the robot is required to be under low gain control during the first two steps. In

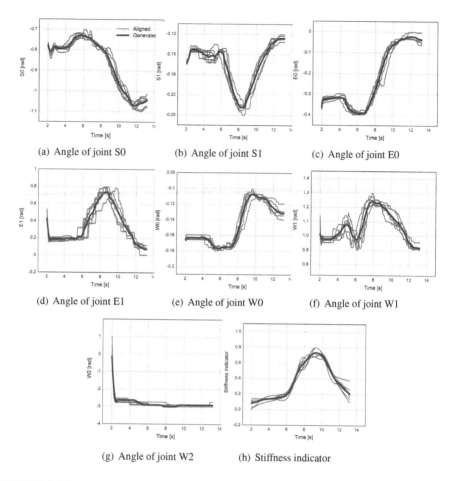

(a) Angle of joint S0   (b) Angle of joint S1   (c) Angle of joint E0

(d) Angle of joint E1   (e) Angle of joint W0   (f) Angle of joint W1

(g) Angle of joint W2   (h) Stiffness indicator

FIGURE 4.24   The result of the alignment of the 7-DoFs movement trajectories (a–g) and the 1-DoF stiffness profile (h).

this case, it is usually unlikely to accomplish the task meeting these two requirements simultaneously, while it can be handed with the proposed framework.

Once the segments of the stiffness profile are obtained, each sequence can be efficiently adapted using the DMP model. To deal with the new task situation, the third and the fourth segments of the stiffness are planned by generalizing the goal of the third sequence, i.e., the initial point of the fourth sequence to a certain value. Fig. 4.25(a) shows three generalized stiffness profiles (red lines) based on the reference one (dark line), namely the generated profile from the last phase. Then, through experience, we select the stiffness profile with a proper goal for the task described above. The position profiles are shown in Fig. 4.24 and their velocities are utilized as the reference profiles in the variable impedance controller.

The new task is performed successfully with the generalized stiffness profile. The collected force profiles of the robot endpoint in z-axis are shown in Fig. 4.25(b). The

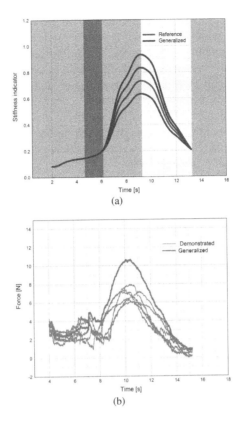

FIGURE 4.25   (a) the planned stiffness profiles; and (b) the measured force profiles in z axis from the six demonstrations (gray lines) and the generalization phase (blue line).

gray lines denote the force profiles measured from the six demonstrations, and the blue line represents one typical force profile obtained from the generalization phase. For better visualization, all the force lines are aligned in the time coordinate. The force profiles and the stiffness profiles have a very similar shape, which means the human-to-robot stiffness transfer has been achieved successfully. More importantly, the force during the generalization phase shows the behavior as expected: keeping low in the first and the second steps and then increasing when the robot lifts the bottle full of water, which can demonstrate the effectiveness of the proposed method.

4.8   CONCLUSION

In this chapter, we introduced a framework for the robot learning and generalizing human-like variable impedance skills that integrate several models for encoding, segmenting, aligning and generalizing the positional trajectories as well as the sEMG-based stiffness profile. By integrating BP-AR-HMM, GTW and DMP into the framework, it allows the potential of our framework to learn a large library of skills including an infinite number of both movement primitives and stiffness primitives, and it also enables to efficiently regulate each individual segment of the demonstrated profiles which can greatly facilitate the generalization of the learned skills. We combine the DMP with GMMs to enable the robot to learn from a set of demonstrations of the same task, which enable robots to learn from multiple demonstrations and generate a better motion trajectory than that just from one demonstration. Additionally, by considering the sEMG-based stiffness estimation of the demonstrator, our proposed framework allows robots to learn variable impedance skills from humans and to generalize the human-like skills to adapt to novel task situations. It is worth noting that this framework can be applied to other robotic platforms, thanks to the model-free characteristics.

The experiments of cutting, lifting water and lift-place tasks show that our method is able to learn and generalize a multi-step task on the Baxter robot manipulator. The experimental results of the pushing tasks have shown that the multi-demonstrations learning method is able to improve the quality of motion skills that the robot has learned and generated. It should also be mentioned that our method can be easily transplanted to other platforms, and this proposed framework can also be applied to a number of potential applications, such as industrial robots, robotic exoskeletons and medical robotic systems, which are required to acquire human-like manipulation skills.

This work makes a preliminary investigation towards the highly effective PbD system, leaving a large number of directions open for the future work such as: (i) a more effective and efficient stiffness estimator will be essential for robots to learn highly human-like features from humans; (ii) some reinforcement learning techniques can be integrated into the framework for the improvement of the DMP policy. In addition, the policy improvement can be effectively achieved by deriving a proper reward/cost function only for the key sub-skills instead of the whole trajectories.

REFERENCES

1. Peidong Liang, Chenguang Yang, Zhijun Li, and Ruifeng Li. Writing skills transfer from human to robot using stiffness extracted from sEMG. pages 19–24, 2015.
2. Emily B Fox, Michael C Hughes, Erik B Sudderth, Michael I Jordan, et al. Joint modeling of multiple time series via the beta process with application to motion capture segmentation. *The Annals of Applied Statistics*, 8(3):1281–1313, 2014.
3. Fanny Ficuciello, Luigi Villani, and Bruno Siciliano. Variable impedance control of redundant manipulators for intuitive human–robot physical interaction. *IEEE Transactions on Robotics*, 31(4):850–863, 2015.
4. Arash Ajoudani, Nikos Tsagarakis, and Antonio Bicchi. Tele-impedance: Teleoperation with impedance regulation using a body–machine interface. *The International Journal of Robotics Research*, 31(13):1642–1656, 2012.

5. Etienne Burdet, Rieko Osu, David W Franklin, Theodore E Milner, and Mitsuo Kawato. The central nervous system stabilizes unstable dynamics by learning optimal impedance. *Nature*, 414(6862):446–449, 2001.

6. Chenguang Yang, Chao Zeng, Peidong Liang, Zhijun Li, Ruifeng Li, and Chun-Yi Su. Interface design of a physical human–robot interaction system for human impedance adaptive skill transfer. *IEEE Transactions on Automation Science and Engineering*, 15(1):329–340, 2017.

7. Chao Zeng, Chenguang Yang, Zhaopeng Chen, and Shi-Lu Dai. Robot learning human stiffness regulation for hybrid manufacture. *Assembly Automation*, 2018.

8. Chenguang Yang, Chao Zeng, Yang Cong, Ning Wang, and Min Wang. A learning framework of adaptive manipulative skills from human to robot. *IEEE Transactions on Industrial Informatics*, 15(2):1153–1161, 2018.

9. Zhao Guo, Yongping Pan, Tairen Sun, Yubing Zhang, and Xiaohui Xiao. Adaptive neural network control of serial variable stiffness actuators. *Complexity*, 2017, 2017.

10. Luís Santos and Rui Cortesão. Computed-torque control for robotic-assisted tele-echography based on perceived stiffness estimation. *IEEE Transactions on Automation Science and Engineering*, 15(3):1337–1354, 2018.

11. Luka Peternel, Tadej Petrič, Erhan Oztop, and Jan Babič. Teaching robots to cooperate with humans in dynamic manipulation tasks based on multi-modal human-in-the-loop approach. *Autonomous robots*, 36(1-2):123–136, 2014.

12. Luka Peternel, Leonel Rozo, Darwin Caldwell, and Arash Ajoudani. A method for derivation of robot task-frame control authority from repeated sensory observations. *IEEE Robotics and Automation Letters*, 2(2):719–726, 2017.

13. Petar Kormushev, Sylvain Calinon, and Darwin G Caldwell. Imitation learning of positional and force skills demonstrated via kinesthetic teaching and haptic input. *Advanced Robotics*, 25(5):581–603, 2011.

14. Auke Jan Ijspeert, Jun Nakanishi, and Stefan Schaal. Movement imitation with nonlinear dynamical systems in humanoid robots. In *Proceedings 2002 IEEE International Conference on Robotics and Automation (Cat. No. 02CH37292)*, volume 2, pages 1398–1403. IEEE, 2002.

15. Jens Kober, Betty Mohler, and Jan Peters. Learning perceptual coupling for motor primitives. In *2008 IEEE/RSJ International Conference on Intelligent Robots and Systems*, pages 834–839. IEEE, 2008.

16. Katharina Muelling, Jens Kober, and Jan Peters. Learning table tennis with a mixture of motor primitives. In *2010 10th IEEE-RAS International Conference on Humanoid Robots*, pages 411–416. IEEE, 2010.

17. Zhijun Li, Ting Zhao, Fei Chen, Yingbai Hu, Chun-Yi Su, and Toshio Fukuda. Reinforcement learning of manipulation and grasping using dynamical movement primitives for a humanoidlike mobile manipulator. *IEEE/ASME Transactions on Mechatronics*, 23(1):121–131, 2017.

18. Franz Steinmetz, Alberto Montebelli, and Ville Kyrki. Simultaneous kinesthetic teaching of positional and force requirements for sequential in-contact tasks. In *2015 IEEE-RAS 15th International Conference on Humanoid Robots (Humanoids)*, pages 202–209. IEEE, 2015.

19. Feifei Bian, Danmei Ren, Ruifeng Li, Peidong Liang, Ke Wang, and Lijun Zhao. An extended DMP framework for robot learning and improving variable stiffness manipulation. *Assembly Automation*, 2019.

20. Yanlong Huang, Fares J Abu-Dakka, João Silvério, and Darwin G Caldwell. Generalized orientation learning in robot task space. In *2019 International Conference on Robotics and Automation (ICRA)*, pages 2531–2537. IEEE, 2019.

21. AR Curtis and JK Reid. The choice of step lengths when using differences to approximate Jacobian matrices. *IMA Journal of Applied Mathematics*, 13(1):121–126, 1974.

22. Man K Kwong and Anton Zettl. *Norm inequalities for derivatives and differences.* Springer, 2006.

23. Feng Zhou and Fernando De la Torre. Generalized time warping for multi-modal alignment of human motion. In *2012 IEEE Conference on Computer Vision and Pattern Recognition*, pages 1282–1289. IEEE, 2012.

24. Hiroaki Sakoe and Seibi Chiba. Dynamic programming algorithm optimization for spoken word recognition. *IEEE transactions on acoustics, speech, and signal processing*, 26(1):43–49, 1978.

25. Chuize Chen, Chenguang Yang, Chao Zeng, Ning Wang, and Zhijun Li. Robot learning from multiple demonstrations with dynamic movement primitive. In *2017 2nd International Conference on Advanced Robotics and Mechatronics (ICARM)*, pages 523–528. IEEE, 2017.

26. Kerstin Dautenhahn and Chrystopher L Nehaniv. The agent-based perspective on imitation, imitation in animals and artifacts. *MIT Press*, 9:105, 2002.

27. Chenguang Yang, Chao Zeng, Cheng Fang, Wei He, and Zhijun Li. A DMPS-based framework for robot learning and generalization of humanlike variable impedance skills. *IEEE/ASME Transactions on Mechatronics*, 23(3):1193–1203, 2018.

28. Sylvain Calinon, Florent D'halluin, Darwin G Caldwell, and Aude G Billard. Handling of multiple constraints and motion alternatives in a robot programming by demonstration framework. In *2009 9th IEEE-RAS International Conference on Humanoid Robots*, pages 582–588. IEEE, 2009.

29. Freek Stulp, Jonas Buchli, Alice Ellmer, Michael Mistry, Evangelos A Theodorou, and Stefan Schaal. Model-free reinforcement learning of impedance control in stochastic environments. *IEEE Transactions on Autonomous Mental Development*, 4(4):330–341, 2012.

30. Klas Kronander, Mohammad Khansari, and Aude Billard. Incremental motion learning with locally modulated dynamical systems. *Robotics and Autonomous Systems*, 70:52–62, 2015.

31. Alexandros Paraschos, Elmar Rueckert, Jan Peters, and Gerhard Neumann. Model-free probabilistic movement primitives for physical interaction. In *2015 IEEE/RSJ International Conference on Intelligent Robots and Systems (IROS)*, pages 2860–2866. IEEE, 2015.

32. Scott Niekum, Sarah Osentoski, George Konidaris, and Andrew G Barto. Learning and generalization of complex tasks from unstructured demonstrations. In *2012 IEEE/RSJ International Conference on Intelligent Robots and Systems*, pages 5239–5246. IEEE, 2012.

33. Mingshan Chi, Yufeng Yao, Yaxin Liu, Yiqian Teng, and Ming Zhong. Learning motion primitives from demonstration. *Advances in Mechanical Engineering*, 9(12):1687814017737260, 2017.

34. Sylvain Calinon. A tutorial on task-parameterized movement learning and retrieval. *Intelligent Service Robotics*, 9(1):1–29, 2016.

5   Learning Human Skills from Multimodal Demonstration

5.1   INTRODUCTION

In order to facilitate physical human-robot interaction and collaboration, a number of light-weight robotic platforms (e.g., Baxter and Sawyer from Rethink Robotics) have been recently developed. These robots are designed and manufactured with the purpose that humans can directly and friendly interact or collaborate with the robots. The output forces/torques are usually restricted into a proper range for the concern of safety. However, it is not easy for a light-weight robot (e.g., Baxter) to perform a task that looks like an easy one (e.g., pushing a button). This may be due to the limited output forces, but more importantly the lack of adaptability which is, however, necessary for the tasks that require the consideration of force/stiffness constraints [11, 6]. This chapter aims to provide a promising LfD solution to this problem.

One advantage of LfD is that human factors are taken into account by integrating the flexibility and adaptability of humans into the human-in-the-loop robot systems [12]. For now, however, a number of problems needed to be addressed before its real applications in industry, one of which should be how to enable robots to perform tasks in a human-like manner, in order to improve the robotic adaptability as stated above. Here, we use the term human-like, referring to that robotic arms share the similarity of the adaptability of human limb muscle control strategies [5] [7]. This work aims to take one step toward this goal by developing a multimodal LfD system. The encoding of multimodal sensor signals has been verified effective in a number of task requirements, e.g., precision motion control [13]. Specifically, the multimodal data considered in this work includes robotic endpoint states (i.e., position and velocity), sEMG signal extracted from human arm, and force data collected from the force sensor mounted between the endpoint and the tool.

In this chapter, we propose a more natural way for the realization of stiffness adaptation by estimating the human tutor's limb stiffness based on the sEMG signals and then transferring the stiffness to the robot arm. The human arm endpoint stiffness can be traced online by using a computationally efficient Cartesian stiffness estimation model [8] [14]. The sEMG-based impedance control strategy has been instigated and successfully applied to a number of robot systems (see, e.g., Refs. [7] [1]). Here, we directly collect the sEMG data from the human tutor's upper limb for stiffness estimation along with kinematics demonstration, without the need for a learning process of obtaining proper stiffness profiles.

For encoding of the demonstration data, DMP is a widely used approach. In Ref. [15], a novel structure of DMP with variable stiffness control is combined to allow synchronization with the demonstrated motion during the operator's intervention. In Ref. [16], a novel coupled movement sequences planning and motion adaption based on DMP is used for a walking exoskeleton robot. In Ref. [17], a temporal coupling for DMP based on a repulsive potential is proposed to keep the desired path shape under velocity constraints. In our work [10] [9], we developed a DMP framework for the representing of the motion and the stiffness simultaneously. However, DMP models each variable separately, without considering the correlation information between different variables. Instead, probabilistic algorithms such as the Hidden Markov Model (HMM) can be used to represent the correlation by encoding a joint-probability density function over the demonstration data. In Ref. [18], Gaussian Mixture Model (GMM) is used to encode the operator's trajectory data. In Ref. [19], a reference-point and object-dependent Gaussian process hidden semi-Markov model (RPOD-GP-HSMM) is used to learn motion primitives. In Ref. [20], a coupled Gaussian process hidden semi-Markov model (GP-HSMM) is used for robots learning rules of interaction. In Ref. [21], GMM and Gaussian Mixture Regression (GMR) are used to share the time of a therapist between multiple patients. In Ref. [22], GMM-GMR is used to analyze the data from human demonstration. In Ref. [23], a hierarchical Dirichlet process-variational autoencoder-Gaussian process-hidden semi-Markov model (HVGH) is proposed to extract features and divide it. In Ref. [24], the HMM-based approach is proposed with the combination of Gaussian Mixture Regression (GMR) to generate the control variables via regression. In Ref. [25], Hidden Semi-Markov Model (HSMM) is further used instead of the HMM model, in order to improve the robustness of the robotic system against external perturbations in temporal space. In Ref. [2], the HSMM-GMR model is further used to model motion as well as force data for in-contact tasks.

Inspired by the work [24] [25] [2], in this chapter we further extend the HSMM model by adding another joint-probability density function for the modeling of the distribution between position and stiffness. This extended information is based on the fact that the demonstration profiles from a human about force, velocity and stiffness, as well as their co-relation with the positions, are all crucial for the robot to learn. Therefore, the learned model is further integrated into an sEMG-based variable impedance control strategy as a unified skill representation model. Thus enabling to integrate sEMG-based variable impedance control strategy into the unified skill representation model. GMR is as well used to generate the expected control variables, which are then properly mapped into an impedance controller.

5.2 SYSTEM DESCRIPTION

The presented approach is shown in Fig. 5.1. It basically consists of two parts: Learning and Reproduction. In the learning part, the collected data are used to estimate the model parameters. And in the reproduction part, the robot performs the same task as demonstrated with the learned knowledge. Specifically, our approach enables the robot to learn task information from the human multimodal demonstration. The experiment result shows that multimodal learning can achieve a better performance

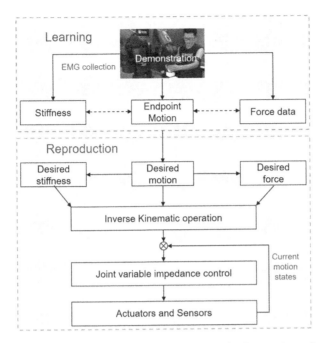

FIGURE 5.1 The overview of the proposed approach. It consists of a two-phase learning-reproduction architecture: in the learning phase, multimodal sensor signals collected from human demonstration are encoded, in order to learn control variables that can account for a specific task. In the reproduction phase, the robot is then required to perform the task based on the learned multimodal control variables.

than using single modality when dealing with a task with position and force constraints like pushing some objects.

The procedure follows three basic phases, including:

i) *Demonstration*: First, the human tutor demonstrates the robot to perform a pushing-pushing task for several times, under the built-in kinematics teaching mode provided by the robot manufacturer. During the demonstration, the robot endpoint posture is recorded, plus the force data and the human arm endpoint stiffness profiles. Note that the human tutor's hands hold on the robot endpoint during the pushing-pushing process.

ii) *Model learning*: Then, the demonstration data (i.e., the set $\{x_i, \dot{x}_i, k, F_i\}$, $i = 1, 2, 3$) are used to fit the HSMM model for the estimation of the model parameters. The orientations are not considered in this work and therefore fixed during the experiment.

iii) *Reproduction*: Finally, the robot reproduces the task under the variable impedance controller with the control variables (position, stiffness, and force) generated by the GMR model.

5.3 HSMM-GMR MODEL DESCRIPTION

Considering a set of observations collected from the demonstrations, i.e., $\{x_m, \dot{x}_m, k_m, F_m\}_{m=1}^{M}$, where x and $\dot{x}$ represent the position and velocity of the robot end-point, respectively; k the robot endpoint stiffness described above, and F the force collected from the force sensor mounted onto the robot endpoint.

First, HSMM is used to model the demonstration data. Then, GMR is used to generate the expected control variables based on the estimated parameters of the HSMM model.

5.3.1 DATA MODELING WITH HSMM

The HSMM model is usually parametrized by:

$$\Theta = \{\{a_{i,j}\}_{j=1, j\neq i}^{K}, \pi_i, \mu_i^D, \Sigma_i^D, \boldsymbol{\mu}_i, \boldsymbol{\Sigma}_i\}_{i=1}^{K} \tag{5.1}$$

where K is the model states. π_i is the initial probability of the ith state. a_{ij} represents the transition probability from state j to i. μ_i^D and Σ_i^D are means and variances, respectively, modeling the K Gaussian parametric duration distributions. $\boldsymbol{\mu}_i$ and $\boldsymbol{\Sigma}_i$ represent mean vectors and covariance matrices of the K joint observation probabilities, respectively.

The ith state duration probability density function is defined as

$$p_i^D(t) = \mathcal{N}(t; \mu_i^D, \Sigma_i^D) \tag{5.2}$$

with $t = 1, \ldots, t_{max}$, where t_{max} is the maximum allowed duration of a HSMM state which can be determined by

$$t_{max} = \gamma \frac{T_{max}}{K} \tag{5.3}$$

where T_{max} is the samples of these demonstrations. γ is a scaling factor that is usually set 2 such that state duration probability density function can be well modeled even if EM converges poorly [25].

The observation probability at each time step t for the ith state is defined by

$$p_i(z_t) = \mathcal{N}(z_t; \boldsymbol{\mu}_i, \boldsymbol{\Sigma}_i) \tag{5.4}$$

with $^1z_t = [\boldsymbol{x}_t^T \ \dot{\boldsymbol{x}}_t^T]^T$, $^2z_t = [\boldsymbol{x}_t^T \ \boldsymbol{k}^T]^T$ and $^3z_t = [\boldsymbol{x}_t^T \ \boldsymbol{F}^T]^T$ are the concatenation of the observed variables at each time step t, corresponding to the three sets of observations, i.e., $\{x_m, \dot{x}_m\}_{m=1}^{M}$, $\{x_m, k_m\}_{m=1}^{M}$, and $\{x_m, F_m\}_{m=1}^{M}$, respectively.

For simplicity, then, the mean vector $\boldsymbol{\mu}_i$ and the covariance matrix $\boldsymbol{\Sigma}_i$ of each of these concatenations are separately represented as

$$\begin{cases} ^1\boldsymbol{\mu}_i = \begin{bmatrix} \mu_i^x \\ \mu_i^{\dot{x}} \end{bmatrix} \\ ^1\boldsymbol{\Sigma}_i = \begin{bmatrix} \Sigma_i^{xx} & \Sigma_i^{x\dot{x}} \\ \Sigma_i^{\dot{x}x} & \Sigma_i^{\dot{x}\dot{x}} \end{bmatrix} \end{cases} \tag{5.5}$$

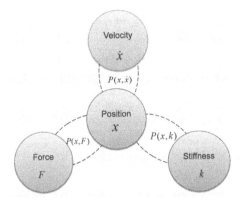

FIGURE 5.2 Graphical representation of the basic idea of the multimodal robotic learning strategy, in which three joint Gaussian distributions are used to encode the multimodal signals. The three HSMM models are trained separately in this work.

and

$$
\begin{cases}
{}^2\boldsymbol{\mu}_i = \begin{bmatrix} \mu_i^x \\ \mu_i^k \end{bmatrix} \\[2ex]
{}^2\boldsymbol{\Sigma}_i = \begin{bmatrix} \Sigma_i^{xx} & \Sigma_i^{xk} \\ \Sigma_i^{kx} & \Sigma_i^{kk} \end{bmatrix}
\end{cases}
\tag{5.6}
$$

and

$$
\begin{cases}
{}^3\boldsymbol{\mu}_i = \begin{bmatrix} \mu_i^x \\ \mu_i^F \end{bmatrix} \\[2ex]
{}^3\boldsymbol{\Sigma}_i = \begin{bmatrix} \Sigma_i^{xx} & \Sigma_i^{xF} \\ \Sigma_i^{Fx} & \Sigma_i^{FF} \end{bmatrix}
\end{cases}
\tag{5.7}
$$

The three sets of $\{\mu_i, \Sigma_i\}$ are used to parametrize the joint Gaussian distributions $\mathscr{P}(x, \dot{x})$, $\mathscr{P}(x, k)$, and $\mathscr{P}(x, F)$, respectively. Namely, these three joint Gaussian distributions are modeled in parallel (see Fig. 5.2). The parameters of the HSMM model, i.e., Θ are estimated based on the demonstration data.

5.3.2 TASK REPRODUCTION WITH GMR

We compute the expected control variables with the GMR model. Their expectations are computed according to the current HSMM state given the reference position:

$$
\dot{\boldsymbol{x}}_t^* = \sum_{i=1}^{K} h_{i,t} [\mu_i^{\dot{x}} + \Sigma_i^{\dot{x}x}(\Sigma_i^{xx})^{-1}(\boldsymbol{x}_t - \mu_i^x)]
\tag{5.8}
$$

$$
\boldsymbol{k}^* = \sum_{i=1}^{K} h_{i,t} [\mu_i^k + \Sigma_i^{kx}(\Sigma_i^{xx})^{-1}(\boldsymbol{x}_t - \mu_i^x)]
\tag{5.9}
$$

$$F_t^* = \sum_{i=1}^{K} h_{i,t}[\mu_i^F + \Sigma_i^{Fx}(\Sigma_i^{xx})^{-1}(x_t - \mu_i^x)] \qquad (5.10)$$

where $h_{i,t}$ represents the brief distribution of the Kth HSMM state. It should be noted that the model parameters $\{\mu_i, \Sigma_i, h_{i,t}\}$ should be different for each corresponding HSMM-GMM model since different modal data have been taken as input into the different models for training. Here, for simplicity, they are written in the same format. And $h_{i,t}$ is computed by

$$h_{i,t} = \mathscr{P}(s_t = i; z_{1:t}) = \frac{a_{i,t}}{\sum_{\kappa=1}^{K} a_{\kappa,t}} \qquad (5.11)$$

The denotation $a_{i,t}$ represents the forward variable of the HSMM model and computed by

$$a_{i,t} = \sum_{j=1}^{K} \sum_{d=1}^{\min(t_{max},t-1)} a_{j,t-d} a_{j,i} p_i^D(d) \prod_{s=t-d+1}^{t} \mathscr{N}(x_s; \mu_i^x, \Sigma_i^{xx}) \qquad (5.12)$$

with initiation in each state:

$$a_{i,1} = \pi_i \mathscr{N}(x_1; \mu_i^x, \Sigma_i^{xx}) \qquad (5.13)$$

where x_1 represents the starting point of the robot endpoint position trajectory. Please see Refs. [24] and [25] for more details of the HSMM-GMR model.

To summarize, once the reference positions x_t and the estimated parameters of the HSMM's states are obtained, the expected velocities, stiffness profiles, and force can be calculated by Eqs. 5.8, 5.9, and Eq. 5.10, respectively.

5.4   IMPEDANCE CONTROLLER IN TASK SPACE

In this work, the robot arm is controlled under the torque control mode with an impedance controller in joint space. Here, we use a commonly used controller, the formation of which consists of four parts:

$$\tau_{cmd} = K_j(q^* - q_{msr}) + D(\dot{q}^* - \dot{q}_{msr}) + J^T F^* + \tau_{dyn}(q, \dot{q}, \ddot{q}) \qquad (5.14)$$

where q^* and q_{msr} are the desired and the measured joint angles during the phase of task reproduction; $\dot{q}^*$ and $\dot{q}_{msr}$ are the desired and the measured joint velocities. K_j and D are the joint stiffness and damping coefficients, respectively. J is the Jacobian matrix of the robot arm. F^* is the desired endpoint force obtained from Eq. 5.10. $\tau_{dyn}(q, \dot{q}, \ddot{q})$ represents the dynamical model of the arm compensating for the forces, i.e., the gravity, the inertia, and the Coriolis forces. The dynamical term can be usually identified by several techniques such as adaptive control and assumed to be known in our work when the robot is controlled under the torque control mode. τ_{cmd} is the generated torque applied to the robot joint. The control diagram is shown in Fig. 5.3.

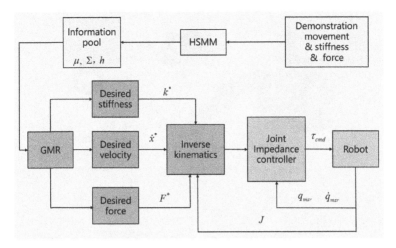

FIGURE 5.3 The HSMM-GMR skill learning diagram. The HSMM model parameters μ, Σ, and h are offline estimated to form the information pool. During the reproduction phase, the desired stiffness, velocity, and force are computed based on the GMR model. The computed variables are then fit into the joint impedance controller through inverse Kinematics. Finally, the computed torques are sent to the robot joints which are enabled under the torque control mode.

The measured joint angles $\boldsymbol{q}_{msr}$ and velocities $\dot{\boldsymbol{q}}_{msr}$ are directly obtained from the interface provided by the robot manufacturer. The desired joint angles $\boldsymbol{q}^*$ are computed through inverse kinematics based on the desired endpoint position. The desired joint velocities $\dot{\boldsymbol{q}}^*$ are computed by:

$$\dot{\boldsymbol{q}}^* = \boldsymbol{J}^+ \dot{\boldsymbol{x}}^* \qquad (5.15)$$

where $\boldsymbol{J}^+$ represents the pseudo-inverse Jacobian matrix, and it is defined by:

$$\boldsymbol{J}^+ = \boldsymbol{J}^T \cdot inv(\boldsymbol{J}\boldsymbol{J}^T + 0.001\boldsymbol{I}) \qquad (5.16)$$

with one unit matrix $\boldsymbol{I}$.

The joint stiffness matrix $\boldsymbol{K}_j$ is computed in accordance with the robot endpoint stiffness:

$$\boldsymbol{K}_j = \boldsymbol{J}^T diag(k_i) \boldsymbol{J} \qquad (5.17)$$

Finally, the joint damping matrix $\boldsymbol{D}$ can be determined by

$$\boldsymbol{D} = diag(d_i) \qquad (5.18)$$

with

$$d_i = \sigma_i \sqrt{k_i} \qquad (5.19)$$

where σ_i are predefined constant coefficients .

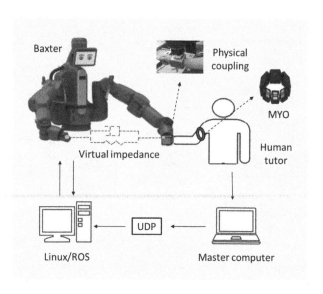

FIGURE 5.4 The experimental setup for skill demonstration.

5.5 EXPERIMENTAL STUDY

5.5.1 BUTTON-PRESSING TASK

An experimental platform based on a Baxter robot is set up for the validation of the proposed method. Fig. 5.4 shows the dual-arm teleoperation system used for skill demonstration.

During the demonstration, the human demonstrator guides the robot to press a button or push a box at a reachable distance from the robot arm. The robot joint state and the human arm muscle sEMG signals are simultaneously recorded for subsequent model training. The changes (e.g., drift) that may appear in sEMG sensors during demonstrations are not considered in this paper because they will not significantly affect the performance of the proposed method. Only several demonstrations are usually needed for most tasks and the long time usage of sEMG sensors will be unnecessary, and the human tutor can control the arm's moving speed and configuration in a proper range.

During the button-pressing experiment, the maximum and minimum joint stiffness of the robot arm joint are respectively set as: $K_r^{max} = [80, 80, 80, 60, 30, 20, 10]$ Nm/rad and $K_r^{min} = [10, 10, 10, 10, 1, 1, 0.5]$ Nm/rad. The constant γ is chosen as 15.

For the sEMG processing, the window size W is set 40 in this work. A set of $M = 6$ demonstrations are obtained and then used to train the HSMM model. The number of states of the HSMM model K is manually chosen as 15 and 20 for the learning of the observed variables $^1z_t = [x_t^T \ \dot{x}_t^T]^T$ and $^2z_t = [x_t^T \ k_{j_t}^T]^T$, respectively.

For comparison, the following four experimental conditions were considered.

Condition 1: force-free control mode. The human demonstrator taught the button-pressing task with the built-in functionality of the robot by grabbing the flange of the robot arm and moving it to approach the button. Then, the robot reproduced the task under the position control mode without involving stiffness regulation.

Condition 2a: the human demonstrated the skill using the dual-arm demonstration for better collection of sEMG signals. The position and velocity control variables (i.e., ^1z_t) were estimated using the HSMM model as described above. The stiffness control variables, however, were learned using dynamic movement primitives (DMP) model from the demonstrated stiffness profiles. Then, the robot reproduced the task under the torque control mode with impedance adaptation.

Condition 2b: the procedure was the same with Condition 2a only with one modification: the stiffness was modeled using the GMM model instead of DMP. Under conditions 2a and 2b, position and stiffness were modeled in a separate manner, which means that the stiffness adaptation is independent of the movement information.

Condition 3: the proposed method was used in this condition. Both the observations were estimated using HSMM. Thus, the correlation between the position and the stiffness can be obtained. The robot was also controlled under the torque control mode with varying impedance.

Condition 4: in order to further test the abilities of our method, we introduced small perturbations into the experiment environment by placing the button 5 mm lower in the z-axis. The experimental procedure is the same as condition 3.

For all the conditions, task reproductions were conducted several times, and no significant variance was obtained between the reproductions under each condition.

Under condition 1, the task's goal could not be achieved. This can be explained by the fact that only position control cannot deal with this force-dominant task which requires stiffness regulation during the physical interaction with the environment.

For conditions 2a–4, the learned joint angle command profiles of joints S0-W1 are shown in Fig. 5.5. The joint W2 is fixed during task demonstration and reproduction for the convenience of mounting the tool. Since the master arm is not used again during the task reproductions, reference trajectories which are estimated from the six demonstrations for these joints are needed in this work. Fig. 5.6 shows the position commands of each joint learned from demonstrations. It shows that our method can generate decent commands, although there are significant variances between the different demonstrations.

TABLE 5.1

The Spearman correlation coefficient (SCC) between the joint angles and the corresponding stiffness profiles.

SCC	Condition 2a	Condition 2b	Condition 3
	0.9102	0.9380	0.9609

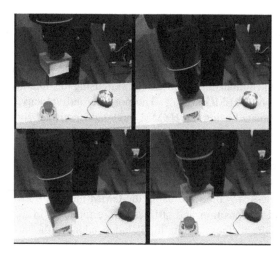

FIGURE 5.5 Successful task reproduction of the button-pressing task. From left to right: the starting pose, moving to the button, pressing the button, and finally leaving the button.

Task reproduction has also not been achieved successfully under conditions 2a and 2b. This can be explained by the fact that the stiffness cannot be modeled well enough by using DMP and GMM compared with the HSMM model. Under conditions 3 and 4, the task has been successfully performed even when there exist small perturbations. An example of successful reproduction is shown in Fig. 5.5. The Spearman correlation coefficient (SCC) between stiffness and position can be coded and increased with the proposed method (see Table 5.1). Thus the dependence of the evolution of the stiffness on position is obtained, resulting in the better performance of representation of the stiffness regulation features.

Take joint S1 as an example. Fig. 5.7 shows the learned stiffness profiles with respect to the demonstrations and the time coordinate. The visual inspection of the lines in Fig. 5.7 suggests that the HSMM model can capture most of the features across the demonstrations. Fig. 5.8 shows the measured position and force profiles of the robot endpoint in z-axis during task reproduction under these conditions. The position and force profiles meet the expectation of the task reproductions regarding the stiffness profiles in Fig. 5.7.

5.5.2 BOX-PUSHING TASK

Another type of task, i.e., the box-pushing task, has also been performed based on the proposed method. In this task, the robot was demonstrated to push a box with a weight of 2.4 Kg placed on the surface of a table along y-axis.

For this task, the maximum and minimum joint stiffness of the robot arm joint are respectively set as: $K_r^{max} = [100, 90, 80, 60, 30, 20, 10]$Nm/rad and $K_r^{min} = [10, 10, 10, 10, 1, 1, 0.5]$Nm/rad.

A set of $M = 5$ demonstrations are obtained and then used to train the HSMM model. The number of states of the HSMM model K is manually chosen as 12 and

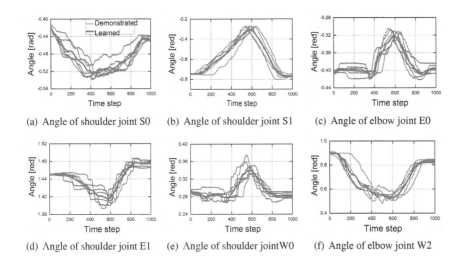

(a) Angle of shoulder joint S0   (b) Angle of shoulder joint S1   (c) Angle of elbow joint E0

(d) Angle of shoulder joint E1   (e) Angle of shoulder jointW0   (f) Angle of elbow joint W2

FIGURE 5.6 The demonstrated and commanded angle profiles of the six joints.

15 for learning the observed variables $^1z_t = [x_t^T \; \dot{x}_t^T]^T$ and $^2z_t = [x_t^T \; k_{jt}^T]^T$, respectively. The parameters for sEMG processing are set the same as in the button-pressing task.

This task has been successfully replayed using our method. It has also been performed several times, and there is no obvious variance observed from these reproductions. The result shows that the stiffness profiles can be well modeled by coding the correlation between them and the position trajectories. As an example, Fig. 5.9 shows the learned stiffness of the joint E1 with respect to the joint angle and time. The SCC between the position and the stiffness of this joint for this task is 0.85. Fig. 5.10 shows the measured position and force profiles of the robot endpoint in y-axis during the task reproduction, which is basically consistent with the demonstrated ones. Table 5.2 shows the RMSE values of the learned stiffness of the joint E1, the measured force and position profiles in the y-axis, which are computed between the demonstrations and reproductions for the ensemble of trajectories.

Note that the proposed method enables the robot to perform the button-pressing and the box-pushing tasks by modeling the stiffness instead of directly modeling the force. Modeling of the force profiles is usually difficult and needs to equip force sensors in robot systems. This experiment suggests that variable stiffness regulation can be used as an impedance modulating strategy for the tasks that do not require precise force control.

5.5.3 PUSHING TASK

The experimental system for human demonstration is shown in Fig. 5.11. During the demonstration, the sEMG signals are collected with the MYO armband and then sent through Bluetooth to a computer for processing. The interaction force signals are collected and amplified by the collection board, then sent to the same computer.

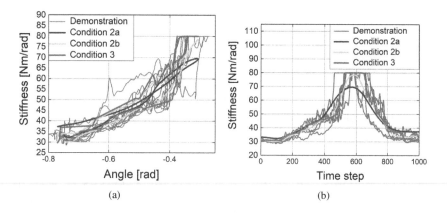

(a)   (b)

FIGURE 5.7   The demonstrated and learned stiffness profiles of joint S1 with respect to (a) joint angle and (b) time step.

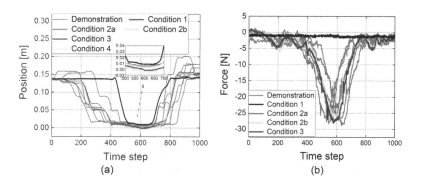

(a)   (b)

FIGURE 5.8   The measured (a) position profiles and (b) force profiles of robot endpoint in z direction during task reproduction.

The processed stiffness and force data are then sent to the host computer through UPD protocol, and then simultaneously recorded along with the robot sate variables. During the reproduction phase, only the host computer is needed, and the generated joint torque commands are directly sent to the robotic joint actuators.

There are a total of 15 sets of demonstration data collected for the pushing-pushing task. The sample rates for the sEMG signals and the force are set as 100 Hz and 200 Hz, respectively. During the reproduction phase, the orientation is fixed as $[\pi, 0, 0]$rad, and the stiffness parameters in orientation are set as $[20, 20, 20]$Nm/rad.

The tasks reproduction is conducted under the following three conditions:

Condition 1: with sEMG and force. The robot reproduces the task with both variable impedance control and force feedforward. In this case, the learned stiffness profiles are used as variable gains in the impedance controller.

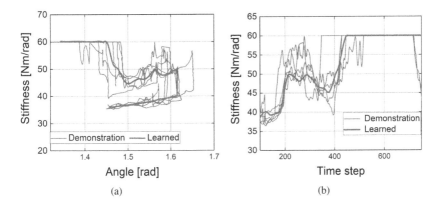

FIGURE 5.9   The demonstrated and learned stiffness profiles of joint E1 with respect to (a) joint angle and (b) time step.

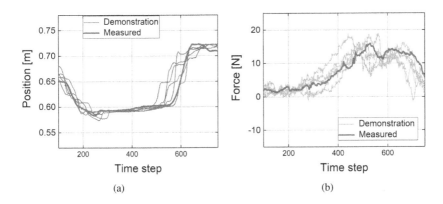

FIGURE 5.10   The measured (a) position profiles and (b) force profiles of robot endpoint in y direction during task reproduction.

Condition 2: with sEMG, without force. The robot performs the task under the variable impedance controller with the learned control variables but without the force term, which means that the robot is controlled under a force-free mode.

Condition 3: with force, without sEMG. The robot reproduces the task with force feedforward and impedance controller but with constant gains. The stiffness parameter in the z-axis is fixed as a constant value.

In summary, as shown in Table 5.3, the first condition is set to validate the proposed control strategy. The second test condition is to show the performance of the control strategy with only sEMG signals, while the third condition is to show that with only force but without the utilization of the sEMG data.

The learned position trajectories and the computed velocity trajectories by GMR model with respect to the demonstration data are shown in Figs. 5.12 and 5.13,

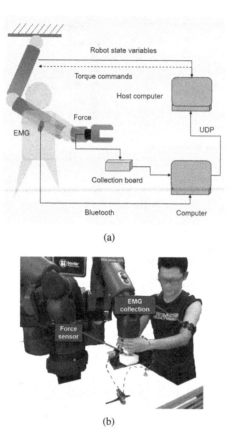

(a)

(b)

FIGURE 5.11   (a) Graphical reinterpretation of the experimental system. During demonstration, multimodal signals (the robot state variables, human arm sEMG signals, and the interaction force between the robot and the environment) are simultaneously collected. All the needed variables are recorded at the same sampling rate on the host computer. (b) The experimental setup during the multimodal demonstration phase. The robot endpoint movement trajectories, the sEMG signals, and the force data are recorded simultaneously during skill demonstration. Note that the first object (stapler) is fixed onto the desk, while the second object (bottle) is placed on the desk without any constraint.

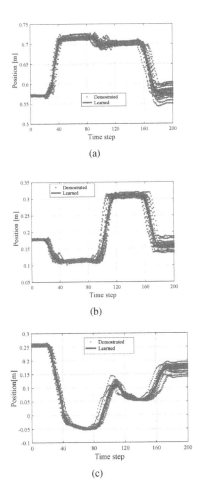

(a)

(b)

(c)

FIGURE 5.12 The position trajectories in the robot end-effector x, y, and z axis respectively. The positions (red lines) are generated based the demonstration ones (gray dots).

TABLE 5.2

The RMS error for the box-pushing task of the profiles with respect to the demonstrations.

RMSE	Position [m]	Stiffness [Nm/rad]	Force [N]
	0.0156	3.230	2.425

TABLE 5.3

The settings for the three test conditions.

Con.	sEMG/Stiffness	Force	Motion traj.
Condition 1	√	√	√
Condition 2	√	×	√
Condition 3	×	√	√

respectively. And the learned velocity profiles with respect to the position trajectories are shown in Fig. 5.13(d)–(f). The computed stiffness in z-axis with respect to the demonstration ones is shown in Fig. 5.14. A typical example of the measured position and force (z-axis) trajectories with respect to the demonstrated ones are shown in Fig. 5.15. The visual inspection shows that the robot achieves the best performance under Condition 1, compared with the other two conditions.

We perform the reproductions for several times under each of these test conditions. Under condition 1, the robot is able to smoothly execute the task, and no obvious variance is observed under this condition. The robot fails to push down the two objects under Condition 2 [see the red lines in Figs. 5.15(b) and (c)], this can be explained by the low force applied onto the robotic endpoint. Under test Condition 3, the task is successfully performed by the robotic arm, even obtaining better performance of position tracking. This is because that the robot manipulator has a better capability of dealing with external perturbations when controlled with high impedance.

Under constant stiffness control mode, however, the robot is not able to reproduce the adaptability of the force to the task situation as demonstrated and as the learned one. A large variance in force is observed under condition 3 (see the black line in Fig. 5.15 as an example). Note that the second object (i.e., the bottle) is not fixed in our experiment. In this case, the rigid robot arm trends to push it away when contacting with the bottle, resulting in the sudden changes of the force profile [see Fig. 5.15(b)]. It suggests that condition 3 may easily cause unstable interactions between the robot and its environment. The force remaining constant from 140 to 160 time-steps may be explained by that there was not enough time for the force sensor to response when the unstable interaction happened. It should be easy to understand that the robot arm with constant high stiffness keeps robust to the external perturbation despite the unstable interaction, and thus the position of the robotic endpoint in the z-direction can almost follow the desired one [see Fig. 5.15(c) 120–160 time-steps].

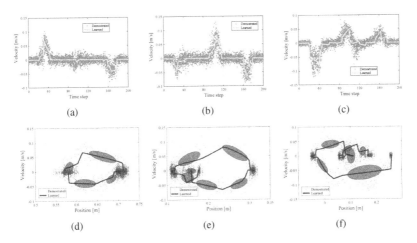

FIGURE 5.13 (a)–(c) show the velocity trajectories in x, y, and z axis respectively. The velocities (red lines) are computed by GMR model based the demonstration ones (gray dots). (d)–(f) are the learned velocity trajectories with respect to the position trajectories with the fitted Gaussian models.

There is a small mismatch in time coordination between the stiffness and the force, which may be explained by the non-synchronously collecting of the sEMG force signals during the demonstration, and by separately training of the stiffness and the force data. Note that the measured force profiles are not only determined by the learned stiffness but also the learned force (see the controller, i.e., Eq. 5.14). Therefore, even in the stable interaction (the fist pushing) the observed robot behavior should be different. We can also see that from 5.15(b) (about the 40th time step) the constant impedance interaction (condition 2) might cause large contact force, which is basically consistent with the experimental findings in Ref. [1].

When humans interact with external environments, we often trend to adapt our arm impedance rather than using a rigid manner [4]. These results show that our method enables robots to learn the features of movement trajectories, stiffness, and force profiles from human demonstration. By encoding multiple sensing signals to extract the correlations between position and the other variables with the combination of HSMM and GMR, our method can generate decent control commands from the demonstration data even the dynamics of these trajectories are complex.

The stiffness and force profiles have demonstrated the features as expected: keep low when getting close to the targets and then increase to push the objects. It should be noted that the adaptation features of the stiffness and force are directly extracted from human demonstration. Multiple information data (movement, stiffness, and force) are included in the proposed teaching-by-demonstration system. In this way, a much more complete skill transfer process can be achieved than only considering one or two signals.

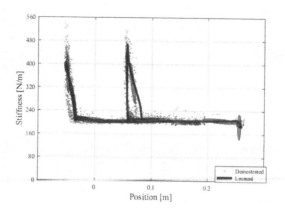

FIGURE 5.14 The stiffness profiles with respect to (a) position and (b) time step in z axis. The red oval areas represent the fitted Gaussian models.

Compared with previous studies (e.g., [2]), our model does not directly encode the "strong" correlation between stiffness and force by separately training the three HSMM models, leaving room for the individually optimizing of impedance and feedforward force, as suggested by the findings in human arm motor learning [3] [4].

One drawback of our approach is that sometimes the models are difficult to train when the demonstrated profiles have a large variance (see Figs. 5.13 and 5.14(b) 40-80 time steps). It may potentially require more demonstration data or larger number of Gaussian models to learn a perfect profile, which would increase the computational cost.

5.5.4 EXPERIMENTAL ANALYSIS

Learning a task by human demonstration such as pressing a button (see Ref. [2]) and pushing an object is sometimes difficult for a lightweight robot. Although the two tasks are quite easy for a human or a traditional heavy-load industrial robot, they are indeed not as easy as expected for a current collaborative robot, e.g., the Baxter robot arm integrated with Series Elastic Actuators (SEAs) as joint actuators. Furthermore, it becomes more difficult when it comes to the learning of the impedance-based skills where both movement and stiffness/force constraints are required to be satisfied simultaneously. Our approach has the capability of addressing this issue by enabling the robot to learn the motor skills, including both movement and stiffness information from the human demonstration.

It should be mentioned that there are other approaches for obtaining variable stiffness profiles. One of them is to derive a stiffness profile based on the force signals by placing a force sensor at the robotic wrist (see, e.g., Ref. [26]). Furthermore, this approach assumes that the stiffness is heavily dependent on the force and should be learned along the force trajectory. In human motor learning, however, it has been validated that the stiffness and the feedforward force are learned separately [3, 4]. Our approach can be extended to simultaneously encode stiffness and force. Some

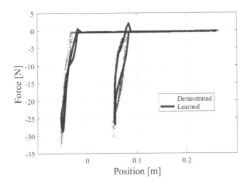

FIGURE 5.15   The force profiles with respect to (a) position and (b) time step in z axis. The red areas in (a) are the fitted Gaussian models. (c) represents the measured potion trajectories of the robotic endpoint in z axis during the task reproduction phase. The blue, red, and black lines correspond to the first, second, and the third experimental conditions, respectively. The sudden changes (black line) are due to the unstable interaction under the third test condition.

tasks may require very delicate force and position control performances, in which cases the dynamics of force need to be well modeled and learned. One possible way to address this is to add another component to consider force information based on the proposed method.

One weakness of our approach is the accuracy of the estimated stiffness since it is difficult to precisely calculate the human arm stiffness so far based on the sEMG signals. Although this is not a problem for most tasks, we will improve our approach to enable the robot to learn more human-like impedance adaptability. The dynamics of the sEMG-based stiffness are often complex (see Figs. 5.7 and 5.9). The stiffness profiles should be more complex in a more complex task situation, in which case it may affect the learning performance; and it would increase computing cost with a larger number of model states and more computing time. Therefore, another direction to improve our approach is to enable effective and efficient learning of stiffness from demonstration data for complex tasks. Furthermore, in this paper, the stiffness is encoded as a diagonal matrix, which may limit the flexibility of the impedance controller in a more complex manipulation task. The complete joint stiffness will be considered in future work, as suggested in Refs. [14, 27].

5.6   CONCLUSION

In this chapter, we propose a teaching-by-demonstration approach considering multiple sensor signals for robots to learn skill features from humans, including movement trajectories, stiffness profiles, and force data. The stiffness profiles are obtained by the estimation of the human tutor's arm impedance based on the collection of the sEMG signals, and the force data are collected from the force sensor rigidly mounted onto the robotic endpoint. These three types of signals are integrated by encoding the three Gaussian distributions, i.e., between position and velocity, stiffness, and force, using the HSMM model. Then, GMR is used to generate the control variables to fit the robotic impedance controller, based on the learned parameters of HSMM. Finally, we demonstrate the validity of the proposed method by a real-word experiment based on the Baxter robot. The experiment suggests compared with the constant stiffness control with only force-sensing involved. The proposed multimodal approach can enable the robot to both successfully and stably perform the pushing task. This method has potential applications in a number of tasks that need both stiffness adaptation and force control.

REFERENCES

1. Arash Ajoudani, Nikos Tsagarakis, and Antonio Bicchi. Tele-impedance: Teleoperation with impedance regulation using a body–machine interface. *The International Journal of Robotics Research*, 31(13):1642–1656, 2012.
2. Mattia Racca, Joni Pajarinen, Alberto Montebelli, and Ville Kyrki. Learning in-contact control strategies from demonstration. In *2016 IEEE/RSJ International Conference on Intelligent Robots and Systems (IROS)*, pages 688–695, Oct 2016.
3. Etienne Burdet, Rieko Osu, David W Franklin, Theodore E Milner, and Mitsuo Kawato. The central nervous system stabilizes unstable dynamics by learning optimal impedance. *Nature*, 414(6862):446–449, 2001.
4. Etienne Burdet, Gowrishankar Ganesh, Chenguang Yang, and Alin Albu-Schäffer. Interaction force, impedance and trajectory adaptation: by humans, for robots. In *Experimental Robotics*, pages 331–345. Springer, 2014.
5. Chenguang Yang, Gowrishankar Ganesh, Sami Haddadin, Sven Parusel, Alin Albu-Schaeffer, and Etienne Burdet. Human-like adaptation of force and impedance in stable and unstable interactions. *IEEE transactions on robotics*, 27(5):918–930, 2011.
6. Yanan Li, Gowrishankar Ganesh, Nathanaël Jarrassé, Sami Haddadin, Alin Albu-Schaeffer, and Etienne Burdet. Force, impedance, and trajectory learning for contact tooling and haptic identification. *IEEE Transactions on Robotics*, 34(5):1170–1182, 2018.
7. Chenguang Yang, Chao Zeng, Peidong Liang, Zhijun Li, Ruifeng Li, and Chun-Yi Su. Interface design of a physical human–robot interaction system for human impedance adaptive skill transfer. *IEEE Transactions on Automation Science and Engineering*, 15(1):329–340, 2017.
8. Arash Ajoudani, Cheng Fang, Nikos G Tsagarakis, and Antonio Bicchi. A reduced-complexity description of arm endpoint stiffness with applications to teleimpedance control. In *2015 IEEE/RSJ International Conference on Intelligent Robots and Systems (IROS)*, pages 1017–1023. IEEE, 2015.
9. Chenguang Yang, Chao Zeng, Yang Cong, Ning Wang, and Min Wang. A learning framework of adaptive manipulative skills from human to robot. *IEEE Transactions on Industrial Informatics*, 15(2):1153–1161, 2018.
10. Chenguang Yang, Chao Zeng, Cheng Fang, Wei He, and Zhijun Li. A DMPS-based framework for robot learning and generalization of humanlike variable impedance skills. *IEEE/ASME Transactions on Mechatronics*, 23(3):1193–1203, 2018.
11. Leonel Rozo, Sylvain Calinon, Darwin G Caldwell, Pablo Jimenez, and Carme Torras. Learning physical collaborative robot behaviors from human demonstrations. *IEEE Transactions on Robotics*, 32(3):513–527, 2016.

12. Luka Peternel, Tadej Petrič, and Jan Babič. Human-in-the-loop approach for teaching robot assembly tasks using impedance control interface. In *2015 IEEE international conference on robotics and automation (ICRA)*, pages 1497–1502. IEEE, 2015.

13. Rui Yang, Poi Voon Er, Zidong Wang, and Kok Kiong Tan. An RBF neural network approach towards precision motion system with selective sensor fusion. *Neurocomputing*, 199:31–39, 2016.

14. Cheng Fang, Arash Ajoudani, Antonio Bicchi, and Nikos G Tsagarakis. Online model based estimation of complete joint stiffness of human arm. *IEEE Robotics and Automation Letters*, 3(1):84–91, 2017.

15. Theodora Kastritsi, Fotios Dimeas, and Zoe Doulgeri. Progressive automation with DMP synchronization and variable stiffness control. *IEEE Robotics and Automation Letters*, 3(4):3789–3796, 2018.

16. Yuxia Yuan, Zhijun Li, Ting Zhao, and Di Gan. DMP-based motion generation for a walking exoskeleton robot using reinforcement learning. *IEEE Transactions on Industrial Electronics*, 67(5):3830–3839, 2019.

17. Albin Dahlin and Yiannis Karayiannidis. Adaptive trajectory generation under velocity constraints using dynamical movement primitives. *IEEE Control Systems Letters*, 4(2):438–443, 2019.

18. Fei Wang, Huan Qi, Yunwen Huang, Xingqun Zhou, Yucheng Long, Xiao Meng, Xiaohan Gao, and Xiaojun Sun. Robot learning by demonstration interaction system based on multiple information. In *2018 IEEE 8th Annual International Conference on CYBER Technology in Automation, Control, and Intelligent Systems (CYBER)*, pages 138–143. IEEE, 2018.

19. Kensuke Iwata, Tatsuya Aoki, Takato Horii, Tomoaki Nakamura, and Takayuki Nagai. Learning and generation of actions from teleoperation for domestic service robots* this work was supported by JST, CREST. In *2018 IEEE/RSJ International Conference on Intelligent Robots and Systems (IROS)*, pages 8184–8191. IEEE, 2018.

20. Satoru Oshikawa, Tomoaki Nakamura, Takayuki Nagai, Masahide Kaneko, Kotaro Funakoshi, Naoto Iwahashi, and Mikio Nakano. Interaction modeling based on segmenting two persons motions using coupled GP-HSMM. In *2018 27th IEEE International Symposium on Robot and Human Interactive Communication (RO-MAN)*, pages 288–293. IEEE, 2018.

21. Carlos Manuel Martinez, Jason Fong, S Farokh Atashzar, and Mahdi Tavakoli. Semi-autonomous robot-assisted cooperative therapy exercises for a therapist's interaction with a patient. In *7th IEEE Global Conference on Signal and Information Processing, GlobalSIP 2019*, page 8969143. Institute of Electrical and Electronics Engineers Inc., 2019.

22. Boyang Ti, Yongsheng Gao, Qiang Li, and Jie Zhao. Dynamic movement primitives for movement generation using GMM-GMR analytical method. In *2019 IEEE 2nd International Conference on Information and Computer Technologies (ICICT)*, pages 250–254. IEEE, 2019.

23. Masatoshi Nagano, Tomoaki Nakamura, Takayuki Nagai, Daichi Mochihashi, Ichiro Kobayashi, and Wataru Takano. High-dimensional motion segmentation by variational autoencoder and Gaussian processes. In *IROS*, pages 105–111, 2019.

24. Sylvain Calinon, Florent D'halluin, Eric L Sauser, Darwin G Caldwell, and Aude G Billard. Learning and reproduction of gestures by imitation. *IEEE Robotics & Automation Magazine*, 17(2):44–54, 2010.

25. Sylvain Calinon, Antonio Pistillo, and Darwin G Caldwell. Encoding the time and space constraints of a task in explicit-duration hidden Markov model. In *2011 IEEE/RSJ International Conference on Intelligent Robots and Systems*, pages 3413–3418. IEEE, 2011.

26. Jianghua Duan, Yongsheng Ou, Sheng Xu, Zhiyang Wang, Ansi Peng, Xinyu Wu, and Wei Feng. Learning compliant manipulation tasks from force demonstrations. In *2018 IEEE International Conference on Cyborg and Bionic Systems (CBS)*, pages 449–454. IEEE, 2018.

27. Arash Ajoudani, Cheng Fang, Nikos Tsagarakis, and Antonio Bicchi. Reduced-complexity representation of the human arm active endpoint stiffness for supervisory control of remote manipulation. *The International Journal of Robotics Research*, 37(1):155–167, 2018.

6 Skill Modeling Based on Extreme Learning Machine

6.1 INTRODUCTION

Human-robot interaction (HRI) plays an increasingly important role in industrial robot application [4, 5]. It is believed that the robot should have the ability to adapt to various demands. Human beings are able to adapt the variation of the surrounding environment, hence, it would be effective if a robot is operated by human depending on their actual abilities and this is defined as teleoperation [6]. The teleoperation innovation with the connection between human and robot has been widely investigated in Ref. [7]. By using robotic teleoperation system, the operator can control a remote robot conveniently through the internet.

Teleoperation system provides a way for robots to imitate the human motion [8]. A direct approach to provide a robot a chance to imitate the human motion has been developed, which is known as motion capture technology and it is a perfect strategy to transfer human abilities to robot side [9]. To realize human motion capture, human body itself ought to be followed first. In the literature, there are various methods to achieve human motion capture. The most widely used method is to estimate the markers from the body of the human operator; however, this may lead to a number of inconveniences to the user. Another method is utilizing image processing from typical cameras. However, this strategy is not reliable owing to the unstable body location capacities during the imaging process. Other methods include stereo-vision cameras that have been applied to motion capture for depth data analysis. Unfortunately, its processing time is quite long, losing effectiveness in real-time applications.

As one of the enabling techniques for teleoperation, motion capture primarily incorporates two interfaces, which are remotely wearable device input interface [10] and detecting interface based on the vision system [11]. Moreover, a few sensors have been utilized for the visual system, such as Leap Motion and Kinect. In Ref.[12], human motions are obtained by a Kinect sensor, and by utilizing the vector approach, people can ascertain the joint points of the Baxter robot. In Ref. [13], the welder-related work is caught by the Leap Motion sensor, which is estimated by a soldering robot with teleoperation networks. Moreover, the wearable devices, for example, exoskeleton [14] or joystick, or Omni haptic device [15] are normally used. In this chapter, we investigate the wearable device, MYO Armband together with the motion capture system using a Kinect sensor to teleoperate a Baxter robot optimized by Kalman filtering-based sensor fusion. In Ref. [16], Kalman filtering (KF) method was used to overcome the shortcomings of Wiener filtering. KF is widely used as it can estimate the past, current, and future state signal, even if the exact nature of the model is not known.

A LfD method consists of a demonstration phase, a learning phase, and a reproduction phase [17]. In the demonstration phase, teleoperation based on HRI attracted much attention [18]. In Ref. [19], a LfD method is presented, which combined haptic feedback and demonstrators variable stiffness to transfer the human physical interactive skill to a robot. To extract the demonstrators' variable stiffness and hand grasping patterns, they collected and processed sEMG signal. In Ref. [20], using an exoskeleton device as the HRI device, they develop a leader-follower teleoperation system. And the system is used in teaching motions of dual-arm robots without a teach pendant. To enhance performance, many techniques or devices are used in HRI, especially visual interaction, which has become one of the most widely utilized techniques.

In the learning phase, neural networks have been widely applied as a training and learning tool [21]. In Ref. [22], they used artificial neural networks (ANN) to recognize gesture patterns and then used a data glove to program a robot. In Ref. [23], they presented an estimating stable dynamical system which mainly based on a neural learning scheme. And the method using ANN can evaluate the system accurately. In this chapter, we propose a robot teaching method which uses a virtual teleoperation system based on visual interaction and uses a neural learning method based on Extreme Learning Machine (ELM). To map human motion on to a robot by HRI, Kinect V2 is used to track human body motion and the hand gesture. And a simulation experiment has been set up based on the V-REP platform where the Baxter robot is guided to learn the demonstrators' motion skills. And a neural learning method-based ELM is developed. It can use a number of training data to approximate a trajectory which is taught by humans to the robot.

In this chapter, the system of teleoperation-based robotic learning is first introduced. Then the human/robot joint angle calculation using Kinect camera, the processing of demonstration data, and the learning method named ELM are presented. Finally, several experiments are used to verify the effectiveness of the proposed methods.

6.2 SYSTEM OF TELEOPERATION-BASED ROBOTIC LEARNING

6.2.1 OVERVIEW OF TELEOPERATION DEMONSTRATION SYSTEM

To delineate the teleoperation of robot utilizing motion capture, an illustrative system was assembled. It comprises of the body tracking system, the Baxter robot, and MYO armbands, as shown in Fig. 6.1.

Motion capture is obtained by the Kinect sensor. In spite of the fact that there are diverse types of Kinect, it is utilized due to its low cost, and it can provide the data required for this research. The Kinect device is associated with a remote computer, wherein processing programming software was introduced and used to get the position information from the Kinect sensor.

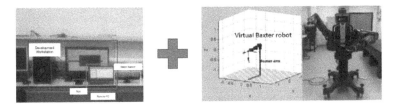

FIGURE 6.1   The experimental teleoperation system, left: development Workstation, MYO armband, Kinect sensor, and Remote PC; right: MATLAB robotic toolbox-based simulated Baxter robot.

This research utilizes Baxter, which is a semi-humanoid robot with arms of 7 DOFs joints and avoidance abilities. Operators can control it through torque, speed, and position mode, respectively. The overall experimental system was associated with and controlled by the development workstation, a remote computer with Ethernet link, and a pair of MYO armbands. The principle of the teleoperation system is represented as Fig. 6.2.

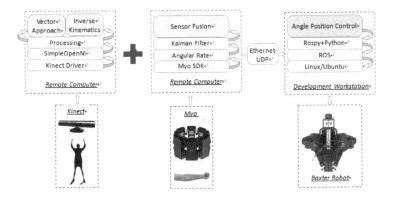

FIGURE 6.2   The experimental teleoperation system, left: development Workstation, MYO armband, Kinect sensor, and Remote PC; right: MATLAB robotic toolbox based simulated Baxter robot.

6.2.2   MOTION CAPTURE APPROACH BASED ON KINECT

(1) General calculation

Commonly, the motion capture calculations for the upper limb depend on distances, locations, and joint angles. The length between two specified points with two and three-dimensional points can be obtained given by Eqs (6.1) and (6.2), respectively.

$$d_{2D} = \sqrt{(x_2 - x_1)^2 + (y_2 - y_1)^2} \tag{6.1}$$

$$d_{3D} = \sqrt{(x_2 - x_1)^2 + (y_2 - y_1)^2 + (z_2 - z_1)^2} \qquad (6.2)$$

where (x_1, y_1) and (x_2, y_2) are points in 2D space, d_{2D} is the distance between these two points, (x_1, y_1, z_1), and (x_2, y_2, z_2) are points in 3D space, d_{3D} is the length between these two points.

The angles at all the joints are obtained by the law of cosines. The most extreme calculable angle is 180 degrees. While computing the angles among the joints, an extra point is required to define at 180~360 degrees. After collecting the motion capture statistics, a triangle is drawn by utilizing any two joint points. From the other two points, the third point of the triangle can be obtained. Under this case, the coordinated statistics for every point of the triangle is known, we are able to find out the length of every side, instead of the value of each angle, which is still unknown. As shown in Fig. 6.3, the magnitude of any coveted point can be calculated by applying the law of cosines.

Computations for the points of joint illustrate the length of sides a, b, c. Similarly, we can also calculate the angles of the triangle using the law of cosines.

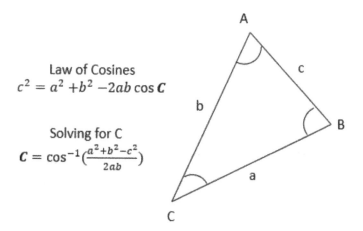

FIGURE 6.3 Mathematical principle description.

(2) Vector approach

Kinect can identify every single joint with coordinate data of the human body and supply with feedback about its statistics. All these directions are transformed into vectors, and the particular angles of the joints can be obtained.

The coordinates of human body joints collected from Kinect in this strategy are under the Cartesian space; additionally, the particular angles from arms are computed. After mapping process by Kinect sensor, they are sent to teleoperate the Baxter as indicated by our requirement.

The five points shoulder Pitch, Yaw, and Roll as well as Elbow Pitch and Roll, shown in Fig. 6.4, are computed from the arm positions data that are extracted from the Kinect.

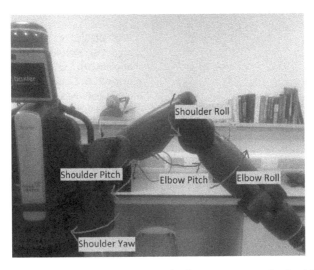

FIGURE 6.4 Demonstration of all related angles in vector approach: shoulder pitch, yaw and roll, Elbow Pitch, and Roll.

The computation of vectors is illustrated in Fig. 6.5. According to Ref. [12], the intense lines CO and CD denote left upper and ahead part of the arm of a human, separately. Intense line BO represents the distance from the left hip to left shoulder, and AO represents the length between right shoulder and left shoulder. The directions with coordinated data B_{X+}, B_{Y+}, and B_{Z+} shows the axis system of Kinect in Cartesian space, where point B is the origin.

Methodology for Computing Shoulder Pitch and Elbow Pitch: As shown in Fig. 6.5, the angle $\angle BOC$ (Shoulder Pitch) is obtained by the distance of two points from vectors $\overline{OB}$ to $\overline{OC}$. The computing methodology be defined by utilizing the three specified joints' position, which are shoulder (point O), elbow (point C), and hip (point B). Delivering these three points using the angle Of() function gives feedback of the value for angles, which are sent to Baxter directly. The $\angle OCD$ (Elbow Pitch), which is the angle among $\overline{CD}$ and $\overline{OC}$, can be computed through sending hand, elbow, and shoulder values into the angle Of() for working as well [12]. In this methodology, we can use the angle Of() command in the Processing software to calculate any angles between two vectors.

Methodology for Computing Shoulder Yaw: As we can see from Fig. 6.5, according to Ref. [12], the angle $\angle EBF$ (Shoulder Yaw) is obtained by a similar method by utilizing both shoulder point and elbow point, which are point A, O, and C, respectively, where the vectors $\overline{OC}$ and $\overline{OA}$ are grouped together. However, the two above mentioned vectors $\overline{OC}$ and $\overline{OA}$ need to be anticipated into the plane XZ. By doing this, we are able to obtain the vectors $\overline{BF}$ and $\overline{BE}$. Angle $\angle EBF$ (Shoulder Yaw) is the value of angle among $\overline{BF}$ and $\overline{BE}$, which can be computed by utilizing angle Of() command in Processing.

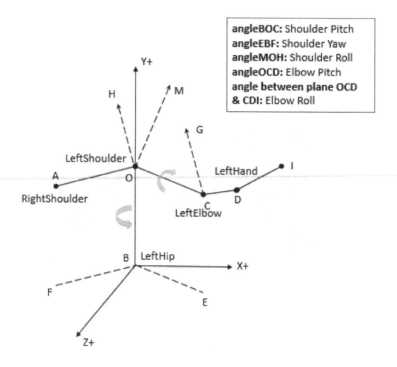

FIGURE 6.5   The principle of vector approach in mathematical computing [12].

Methodology for Computing Elbow Roll and Shoulder Roll: The Elbow Roll is the angle between plane OCD and CDI, which can be calculated by the angle of function. According to angular estimations, Shoulder Roll is most difficult to be calculated. As the computing is not straightforward and all the points are given in the 3D plane, hence, the similar computing method utilized above cannot be accessible here.

The point made by the vectors from elbow to hand, where its plane is opposite to the one from shoulder to elbow. It is going through the shoulder joint as well. The reference vector must be stably concerning the body. As a result, the reference vector can be computed by the intersecting point of vectors from shoulder to shoulder and shoulder to elbow.

For this situation, according to, the normal line from that intersecting point of two vectors is used for continuous calculation. The vector $\overline{OM}$ can be obtained by verifying the intersecting point between vectors $\overline{OC}$ and $\overline{OA}$. The vector $\overline{OM}$ is vertical to the plane obtained from the vectors $\overline{OA}$ and $\overline{OC}$. Clearly, vector $\overline{OM}$ and vector $\overline{OC}$ are vertical from each other.

Along these lines, the normal line vector $\overline{CG}$ can be decided via intersecting point of vectors $\overline{CD}$ and $\overline{OC}$, which is additionally vertical to vector $\overline{OC}$. At this point, we can obtain the vector $\overline{OH}$ by deciphering vector $\overline{CG}$ along the vector $\overline{CO}$ to point O.

The angle $\angle MOH$ between vectors $\overline{OH}$ and $\overline{OM}$ is defined as Shoulder Roll [12].

The orientation angles sent by the Kinect can be separated by utilizing PMatrix3D with Processing software. The PMatrix3D outputs the required rotation matrix as well, where the current coordination framework is well given the backup into the stack. It is then delivered to the shoulder joint. Additionally, the rotation matrix is utilized to transform into the coordination system data. Every single computation in this capacity will be decided within the obtained coordination framework.

After the computation of Shoulder Roll and Elbow Roll angles, the rotation matrix from the stack can be recovered to obtain the initial coordination framework. The right Shoulder Roll is additionally computed with a similar method. Furthermore, a small change has been applied to the vectors coordination system.

As the function used to calculate roll angles is not accurate, the error needs to be corrected. Every value changes of Shoulder Roll is along with the value changes of Shoulder Yaw. The statistics are plotted using MATLAB, as seen in Fig. 6.6. From several trials, the error is mostly revised by the equation demonstrated as follows:

$$\gamma_s = -\gamma_s - \beta_s/2 - 0.6 \tag{6.3}$$

where we define the angle of left shoulder roll is γ_s, and the angle of left shoulder yaw is β_s. These returned values of angles are sent to Baxter development platform for further advanced work utilizing UDP protocol. The data packets created by the server is sent through the function introduced above. So far, every single angular value is sent to teleoperate the Baxter robot with the Python script based on the KF sensor fusion.

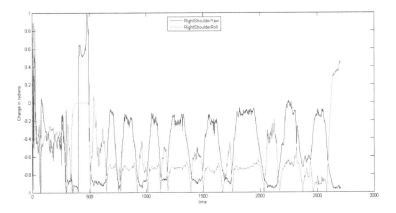

FIGURE 6.6   Error of vector approach.

6.2.3   MEASUREMENT OF ANGULAR VELOCITY BY MYO ARMBAND

The joint angles are obtained by computing the integral of angular velocity. Any positions of human operator's arms can be used as the initial position, where the joint angles are assumed to be zero, according to Ref. [3]. When the operator moves his arm to a new pose P, the rotation angles (joint angles) are the pose P [3].

The frame (X_1, Y_1, Z_1) represents the orientation of MYO armband in the initial position. The frame (X_2, Y_2, Z_2) represents the current orientation of the MYO. From the first MYO armband worn on the upper arm, we can obtain three angles' angular velocity $v1_x$, $v1_y$, $v1_z$, which represent shoulder roll, pitch and yaw, respectively.

From the second MYO armband worn on the forearm, we can get the angles' velocity $v2_x$, $v2_y$, which represent elbow roll and pitch.

Existing in the joint angular velocity measured by the joint angle, there will be errors, however here the errors will be superimposed. The shoulder joint error will be superimposed on the elbow joint, resulting in a greater elbow error. In addition, the integration time problem also leads to the existence of errors. Although the sampling frequency of IMU in the MYO is 50 Hz, the resulting angle will have a large difference in value when the joint angle is calculated from the angular velocity integral in the program in Ref. [3]. Here, the method for the angular velocity was extended from the method of measurement for the angles using MYO armbands mentioned in previous research [3]. In summary, in this chapter, MYO armband is used to measure the angular velocity of each joint, and Kinect is used to get the angles of each joint.

6.2.4   COMMUNICATION BETWEEN KINECT AND V-REP

The communication structure of the virtual teleoperation system is shown in Fig. 6.7. To connect the computer and Kinect V2, we use the Kinect SDK 2.0 for Windows. And we use remote API in the system, where a custom C++ application is used as the client-side and a V-REP scene is server-side. To connect the C++ application and V-REP, one command needs to be put into them, respectively.

The code in C++ project:

clientID = simxStart((simxChar*)"127.0.0.1", 19999, true, true, 2000, 5);

The script in V-REP:

simExtRemoteApiStart(19999);

To prevent causing a delay when controlling the robot by Kinect, we use the non-blocking function call to send the joint angle to the robot. At the same time, the C++ application records the data of the joint angle and save it in a file. A sample command is shown below:

simxSetJointTargetPosition(clientID, jointhandle,jointPosition,

simxOpmodeOneshot);

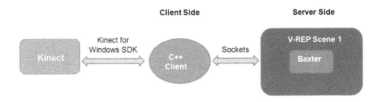

FIGURE 6.7 Error of vector approach.

6.3 HUMAN/ROBOT JOINT ANGLE CALCULATION USING KINECT CAMERA

In order to estimate the 7-DOF model of the human arm, the Denavit-Hartenberg (D-H) coordinate framework has been created in Fig.6.8 with D-H parameters of the kinematic model (human arm) in section 6.1.

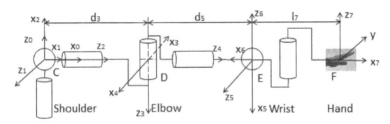

FIGURE 6.8 The D-H coordinate framework and the initial position for each joint of human arm.

TABLE 6.1
Model representation of the DH parameter table [24]

LinkNumber	θ_i	$d_i(m)$	$a_i(m)$	$\alpha_i(rad)$
1	θ_1	0	0	$\pi/2$
2	θ_2	0	0	$\pi/2$
3	θ_3	d_3	0	$\pi/2$
4	θ_4	0	0	$\pi/2$
5	θ_5	d_5	0	$\pi/2$
6	θ_6	0	0	$\pi/2$
7	θ_7	0	a_7	0

Kinect emits infrared rays and detects infrared light reflections so that the depth values of each pixel in the field of view can be calculated, which is the depth data. Wherein, the object body and shape are first extracted. Then using the information above, the position of each joint can be obtained shown in Fig. 6.9.

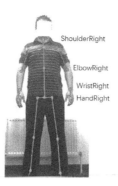

FIGURE 6.9   Image of body skeleton captured by Kinect.

To obtain the angle rotation angle of the human shoulder, we can use the joints of ShoulderRight and ElbowRight in 3D coordinates (x, y, z) to calculate. Assuming three-dimensional coordinates of ElbowRight is (x_1, y_1, z_1), and the two skeletal nodes in the three-dimensional space to form a straight line (l_1). As the shoulder joint is unchanged during the rotation in the z-coordinate, there is just the xoy plane need to be considered. The linear equation is given below:

$$y = k_1 x + b_1 \tag{6.4}$$

where $k_1 = tan\theta_1 = \frac{|y_2 - y_1|}{x_2 - x_1} (x_1 \neq x_2)$, $b1$ is not given in the calculation of the angle of rotation, so the formula is not given here. Assuming the angle between l_1 and y axis is θ_1, as shown in Fig. 6.10, θ_1 is the human shoulder rotation angle, the formula is:

$$\theta_1 = arctank_1 = arctan\left(\frac{y_2 - y_1}{x_2 - x_1}\right) \tag{6.5}$$

FIGURE 6.10   Rotation angel of shoulder joint.

To obtain the angle of rotation of the human elbow, we can use joints of ElbowRight and WristRight in 3D coordinates to work out. Assuming the 3D coordinates of WristRight is (x_2, y_2, z_2), similarly, there is a straight line constituted by

ElBowRight and WristRight (l_2), the linear equation is given below:

$$y = k_2 x + b_2 \tag{6.6}$$

where $k_2 = tan\theta_2 = \frac{|y_3 - y_2|}{x_3 - x_2}$ ($x_2 \neq x_3$), $b2$ is not given in the calculation of the angle of rotation, so the formula is not given here. Assuming the angle between l_2 and l_1 is θ_2, as shown in Fig. 6.11, θ_2 is the human elbow rotation angle, the formula is:

$$\theta_2 = arctank_2 = arctan\left(\frac{k_1 - k_2}{1 + k_1 k_2}\right) \tag{6.7}$$

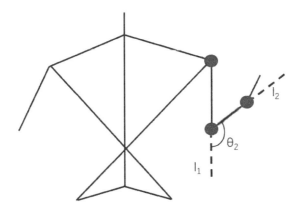

FIGURE 6.11   Rotation angel of elbow joint.

Similarly, for the angle of rotation of the human wrist, which is the angle between l_3 and l_2. This is defined as θ_3 and shown in Fig. 6.12, and the formula is:

$$\theta_3 = arctank_3 = arctan\left(\frac{k_2 - k_3}{1 + k_2 k_2}\right) \tag{6.8}$$

6.4   PROCESSING OF DEMONSTRATION DATA

6.4.1   DYNAMIC TIME WARPING

DTW is an effective time-series matching method for different lengths and is widely used in the field of time series processing and signal processing [25]. Earlier applications were in areas such as speech recognition. Given a candidate area sample C with a width of M and a character sample Q to be queried with a width N, the size of the candidate area is the same as the size of the character to be queried, $M=N$.

Considering that the writing task is well structured and continuous, for any two columns where there is no cross-matching occurring in between and each column

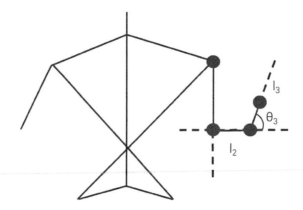

FIGURE 6.12   Rotation angel of wrist joint.

can find the other for matching, which results in the same continuity [26]. The restriction ensures that the i−th column of the sample C and the j−th column of the sample Q have an accumulated distance $D(c_i, q_j)$, and it is only jointly determined by $D(c_i, q_{j-1})$, $D(c_{i-1}, q_j)$, $D(c_{i-1}, q_{j-1})$, and $d(c_i, q_j)$, which is shown as follows [25]:

$$D(c_i, q_j) = min \left\{ \begin{array}{l} D(c_i, q_{j-1}) \\ D(c_{i-1}, q_j) \\ D(c_{i-1}, q_{j-1}) \end{array} \right\} + d(c_i, q_j) \qquad (6.9)$$

where both i, $j > 1$, $d(c_i, q_j)$ is the distance function between two samples; in this chapter, there are five-demonstrations. Hence there are also five-dimensional features f_0 - f_4, and the distance between element c_i in column i in C and element q_j in column j in Q, is defined by the Euclidean distance defined as [25]:

$$d(c_i, q_j) = \sum_{k=0}^{4} (c_{if_k} - q_{jf_k}) \qquad (6.10)$$

where c_{if_k} is the k−th dimension of the i−th column in sample C, q_{jf_k} is the k−th dimension of the j−th column in sample Q.

Although the same tasks differ from execution to execution, this difference should be kept within a small local area. Therefore, the global path must be constrained to maintain the invariance of the local structure and accelerate the solution of the problem. It is important to limit the spatial distance between the two columns c_i and q_j to be less than r elements [27].

$$\|i - j\| \leq r \tag{6.11}$$

$$r = \lceil k * seqL \rceil \tag{6.12}$$

where k is a constant coefficient, and $seqL$ is the length of the feature sequence.

Using the dynamic programming method can accelerate the solution of the horizontal distance $D(c_M, q_N)$. Since the feature sequence of the character Q to be queried and the feature sequence length of the candidate domain are the same, and it is not necessary to normalize the accumulated distance $D(c_M, q_N)$ by the sequence length [27]. Considering the length and width of the characters are inconsistent, and ultimately contribute to inconsistency, hence in this paper, we use the weighted two-way DTW algorithm to calculate the final distance of two characters, the weight of the length of the sequence itself:

$$dist(C, Q) = D(c_N, q_N) * N + D(c_M, q_M) * M \tag{6.13}$$

Finally, each candidate sample C and the dis-queried character Q are arranged in ascending order from $dist(C, Q)$ to obtain a list of candidate regions. Since each character will have several key points, the overlapping candidate fields need to be eliminated. According to the obtained candidate sample list, for any single sample inside, if there is an order prior to and if the overlapping area ratio exceeds the threshold, it is eliminated. The final list is the final test result.

6.4.2 KALMAN FILTER

Kalman filter is an useful linear filtering approach to solve the criterion of mean square error, which estimates the current value of the signal based on the previous estimate and the last observation data. It is estimated by the state equation recursively, and its solution is given in the form of an estimate of its signal model, which is derived from the state equation and the measurement equation as shown below [2]

$$\begin{aligned} \dot{x}(t) &= Ax(t) + Bu(t) + G\omega(t) \\ y(t) &= Hx(t) + v(t) \end{aligned} \tag{6.14}$$

where x and u are state variables; y is the measurement vector; A is the system matrix; G and B are voice matrixes; H is the measurement matrix; ω is the white noise vector; v is the continuous value of the white noise vector. The continuous-time KF updating equation, according to Ref. [2] is demonstrated in (6.15),

$$\begin{aligned} \dot{\hat{x}}(t) &= A\hat{x}(t) + Bu(t) + K[y(t) - H\hat{x}] \\ K(t) &= P(t)H^T r^{-1}(t) \\ \dot{P}(t) &= P(t)H^T + AP(t) - P(t)H^t r^{-1} HP(t) \\ &\quad + Gs(t)G^T \end{aligned} \tag{6.15}$$

where K is the filter gain matrix, $\hat{x}$ is the estimated value of x, and P is the estimated covariance matrix.

The signal and noise in the Kalman filter are represented by the state equation and the measurement equation. The design of the Kalman filter, therefore, requires a known state equation and a measurement equation. It can be used for both smooth and unstable random process, but can also be applied to solve non-time-varying and time-varying systems. We assume that every single joint of human arms is taken into account separately to research, which gives that all the KF factors are the first order, hence, here $A=0$, $B=1$, $G=1$, and $H=1$. Then the KF equations are simplified as below,

$$\dot{x}_i = u_i + \omega$$
$$y_i = x_i + v \tag{6.16}$$

where according to Ref. [2], in this special case, y_i is the angular position of the number of i joint collected from the Kinect sensor, we give it conception as following: $y_i = q_{di}'$. And u_i is the angular velocity of the number of i joint of the operator's arm motion.

$$\dot{\hat{x}}_i = u_i + k\left(y_i - \hat{x}_i\right)$$
$$k = pr^{-1} \tag{6.17}$$
$$\dot{p} = p - pr^{-1}p + s$$

where k is the filter gain matrix, p is the estimated covariance matrix, $\hat{x}_i$ is the required (satisfied) data obtained from KF-based sensor fusion.

6.4.3 DRAGON NATURALLY SPEAKING SYSTEM FOR VERBAL COMMAND

Dragon NaturallySpeaking is a speech recognition software released by Dragon Systems of Newton, Massachusetts [28]. By using Dragon NaturallySpeaking, the operators are able to create documents, reports, emails, fill in forms, and workflows with verbal command [28]. By speaking to the computer, the words appear as text in Microsoft Office Suite, Corel WordPerfect, and all Windows-based applications. Significantly, operators are able to create voice commands to make the computer run applications in multiple steps, which is time-saving.

Dragon NaturallySpeaking software will be used to transfer the voice data to text data and to generate robot motion control commands in our experiments, as shown in Fig. 6.13.

6.5 SKILL MODELING USING EXTREME LEARNING MACHINE

Using the virtual teleoperation system, we can control the Baxter to track a desired to a trajectory. However, in the process of accomplishing a task, human action is not necessarily optimal. Therefore, controlled by teleoperation, the robots' trajectory will have some deviation compared with the target trajectory. Through the neural

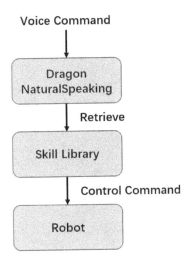

FIGURE 6.13 Dragon NaturalSpeaking software is used to generate robot motion control commands.

network learning based on ELM, the robot trajectory can approach the target trajectory. The dynamical system can be expressed by a first-order autonomous ordinary differential equation

$$\dot{s} = f(s) + \varepsilon \tag{6.18}$$

where s denotes the robot's joint angles, and $\dot{s}$ is the first derivative of s. The dataset is $\{s,\dot{s}\}_{t=0}^{T_1,...T_L}$, ε is a zero mean Gaussian noise [1]. The goal is to obtain an estimation of $\hat{f}$ from f.

To achieve this goal, we use a method based on ELM, which is more efficient than the traditional learning algorithm under the same conditions. To use the ELM in the teleoperation system, the goal is to learn the mapping $f : s \rightarrow \dot{s}$ based on the dataset $\{s,\dot{s}\}_{t=0}^{T_1,...T_L}$.

As shown in Fig. 6.14, for a neural network with a hidden layer, the input layer has n nodes, which is the dimension of s. In the hidden layer, the target function is,

$$\dot{s} = \sum_{i=1}^{L} \beta_i f_i(s) = \sum_{i=1}^{L} \beta_i g\left(\omega_i^T s + b_i\right) \tag{6.19}$$

where g is activation function, $W = (\omega_1, \omega_2, \ldots, \omega_L)^T$ is the input weights, which has dimension $L \times d$; and $\beta = (\beta_1, \beta_2, \ldots, \beta_L)^T$ is the output weights, which also has dimension $L \times d$; $b = (b_1, b_2, \ldots, b_L)$ is the hidden layer biases, $\omega_i^T s$ is inner product of W and s.

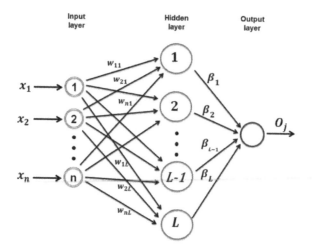

FIGURE 6.14 Extreme learning machine.

The learning goal of a single hidden layer neural network is to minimize the output error. So ELM solves the problems as follows:

$$\min_{\beta} \|H\beta - 0\| \tag{6.20}$$

where

$$H(\omega_1, \omega_2, \ldots, \omega_L, b_1, b_2, \ldots, b_L, s_1, s_2, \ldots, s_L)$$

$$= \begin{bmatrix} g(\omega_1^T s_1 + b_1) & \cdots & g(\omega_L^T s_1 + b_L) \\ \vdots & \ddots & \vdots \\ g(\omega_1^T s_L + b_1) & \cdots & g(\omega_L^T s_L + b_L) \end{bmatrix} \tag{6.21}$$

is the output of the hidden layer node, $O = (\; o_1, \quad o_2, \quad \ldots, o_L \;)^T$ is expected output. In the system, O is the target value which is generated by demonstration.

Once the input weights and hidden layer biases are fixed, the output matrix of the hidden layer H is uniquely determined. Then the problem about training single hidden layer neural network can be transferred into a problem about solving a linear system. The solution is

$$\beta = H^+ O \tag{6.22}$$

where H^+ is the Moore–Penrose generalized inverse of the matrix H.

For any $t \in R$, the activation functions $g(t)$ should be continuous and continuously differentiable. We use a kind of activation functions which satisfy that

$$\begin{aligned} g(t) &= 0, \quad t = 0 \\ g(t) &> 0, \quad \forall t \in R \end{aligned} \tag{6.23}$$

In order to satisfy the above properties, a bipolar sigmoid function

$$g(t) = \frac{2}{1 + e^{-t}} - 1 \tag{6.24}$$

is used. The sigmoid functions are continuous and continuously differentiable, so they have not an impact on the performance of ELM.

6.6 EXPERIMENTAL STUDY

6.6.1 MOTION CAPTURE FOR TRACKING OF HUMAN ARM POSE

The indoor experiment environment is of sufficient illumination, so the Kinect can work well. One operator stands in front of Kinect at a distance of about 2 meters. As we demonstrated in the previous section, we choose the Shoulder Pitch, Shoulder Yaw, Shoulder Roll, Elbow Pitch, and Elbow Roll to apply in this experiment. Following the experimental data collection, a simulated Baxter robot in Matlab was used to teleoperate with the operators.

An operator wore a pair of MYO armbands faces to the Kinect sensor (as seen in Fig. 6.15) with different arm movements. The operator wore one MYO armband near the center of the upper arm and wore the other near the center of the forearm. The former measures the orientation and angular velocity of Shoulder Pitch, Yaw, and Roll. The latter predicts the orientation and the angular velocity of Elbow Pitch and Roll. Before the experiments, it is significant to calibrate the MYO armband and warm up the EMG sensors, in order to recognize different hand postures better for the MYO armband. The operator should not even more any short distances, under this case, only both arms of the operator can freely move with a low and stable speed.

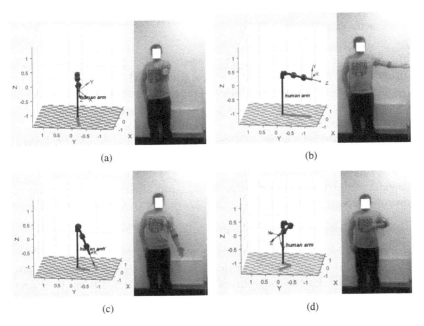

(a)   (b)

(c)   (d)

FIGURE 6.15 Demonstration of experiment at the different positions.

Fig. 6.16 illustrates the graphical results of the 5 selected DOFs with different trajectory after the KF-based sensor fusion between Baxter and the operator. The Kinect sensor gives us the position difference between the robot's real trajectory and the base point and the MYO armbands gives us the angular velocity of those 5 angles accordingly. Then they were fused together via KF. Here, the experimental data of the operator's arm motion from Kinect and MYO, the optimum output from Kalman filter-based sensor fusion and the angular statistics of the simulated Baxter robot was taken, respectively for the test. From the graph, it is concluded that the total were performance of motion capture system becomes improved by applying the KF-based sensor fusion.

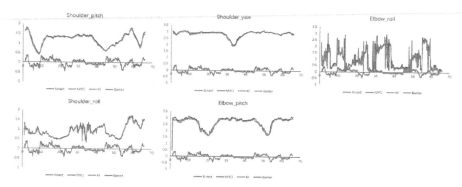

FIGURE 6.16   Controlled by the human with a Kinect, the Baxter can push down the building block on the desk.

The method of using Kinect and MYO armband after KF-based sensor fusion to teleoperate a Baxter robot was developed and validated. The experimental results shown in Table 6.2 demonstrate a series of ratios, which are the different values of the 5 angles between those obtained by KF and those directly collected by Kinect. Because the values obtained via KF are optimum, and the noises during the teleoperation process have been removed. Hence, that ratio is approximately defined as the efficiency improvement, which is denoted as r_e as defined in (6.25) below. The ratios shown in Table 6.1 averagely at 3.185%, 4.086%, 3.442%, 3.269%, and 3.673% for the angular positions of shoulder pitch, shoulder roll, shoulder yaw, elbow pitch, and elbow roll, respectively.

$$r_e = \frac{p_{KF} - p_{Kinect}}{p_{Kinect}} \tag{6.25}$$

where p_{KF} and p_{Kinect} are the experimental data of different angular positions obtained from the KF-based sensor fusion and directly collected from the Kinect, respectively.

6.6.2   TELEOPERATION-BASED DEMONSTRATION IN VREP

Our goal is to explore whether a robot can accomplish a task after teaching by teleoperation, which is based on visual interaction and ELM. To verify the effectiveness

TABLE 6.2
Table for efficient improvement of different angular positions

Data	Shoulder Pitch	Shoulder Roll	Shoulder Yaw	Elbow Pitch	Elbow Roll
Ratio	3.185%	4.086%	3.442%	3.269%	3.673%

of this TbD method, a simulation scene is designed in V-REP. And the scene consists of a Baxter robot, a desk, and some rectangular building blocks. As shown in Fig. 6.17, using the Kinect, a human demonstrator control the Baxter robot to pull down a building block, and then the robot's arm returns to its original position. This action will be done many times. At the same time, the robot joint angle in this process is recorded at regular intervals.

FIGURE 6.17 Controlled by the human with a Kinect, the Baxter can push down the building block on the desk.

Because each simulation time we control the robot is different, the number of data recorded in the experiment are also different. We randomly interpolate the experimental data. By using these processed joint angle to control the robot, the robot can reproduce the same trajectory, which confirms that this method has no effect on the effect of the robot's trajectory. Therefore, a sample with a dimension of 142×3 is got from each simulation after data processing.

We implement the learning algorithm in MATLAB. The function in MATLAB is mainly used to build the ELM. During the training process, the input data is time t_i, and the output data is the joint angle q_i ($i = 1, 2, ...$). The relationship between the mean square error (MSE) and the number of hidden neurons is shown in Table 6.3. Using the constructed ELM to approximate the robot trajectory, we get three groups of output data. As shown in Fig. 6.18, they are the values of the robot's corresponding joint angles, respectively.

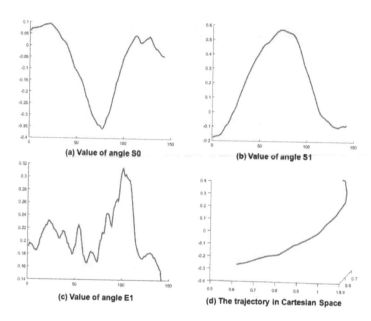

(a) Value of angle S0

(b) Value of angle S1

(c) Value of angle E1

(d) The trajectory in Cartesian Space

FIGURE 6.18   The output data of the RBF network. (a), (b), (c) are the joint angles of S0, S1, and E1. And (d) is the trajectory of the Baxter arm moving downward in Cartesian space.

TABLE 6.3
THE RELATIONSHIP MSE AND THE NUMBER OF HIDDEN NEURONS

Neurons	50	100	150	200	250	300
MSE(S0)	0.0122	0.0088	0.0081	0.0079	0.0079	0.0079
MSE(S1)	0.0391	0.0183	0.0117	0.0107	0.0106	0.0106
MSE(S2)	0.0073	0.0072	0.0071	0.0070	0.0070	0.0070

In the next step, the data is sent to V-REP through MATLAB to control the Baxter robot. As shown in Fig. 6.19, the Baxter in V-REP can autonomously complete the task which is taught by a human Demonstrator.

6.6.3   VR-BASED TELEOPERATION FOR TASK DEMONSTRATION

To verify the effectiveness of the robot teaching system, we designed a task in which the operator should control the robot to draw points in the circles. The system structure is shown in Fig. 6.20. As shown in Fig. 6.21, controlled by LEAP Motion controller, the robot moved from its initial position, drawn a point in the first circle,

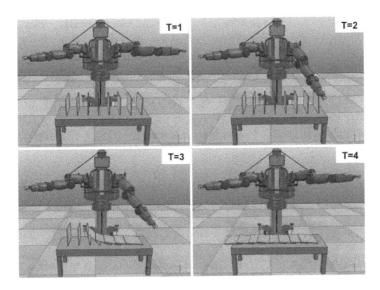

FIGURE 6.19 The experiment results 3: Through learning and training by RBF network, the Baxter can push down the building block autonomously.

then moved to draw a point in another circle, and finally returned to the endpoint. At the same time, the scene of the robot and its work platform was transmitted to the Unity and the Oculus DK 2. The scene seen through the Oculus DK 2 seems to be in front of the operator, so the operator can control the robot more directly and more naturally. Moreover, the delay of the entire system is negligible.

And this demonstration was repeated 8 times. At the same time, we recorded the position coordinates of the robot in the Cartesian space. Because each demonstration is different, there are some differences between each set of data. As shown in Fig. 6.22, the changes in the z-coordinate of robot end-effector position are different in each demonstration. We need to align them on the timeline. In this process, the Dynamic Time Warping (DTW) algorithm is used to make it.

As shown in Fig. 6.23, processed by the DTW algorithm, the data of Z-coordinate has the same dimensions, and they are aligned in time. The data of X coordinate and Y coordinate also went through the same process. So we got 8 sets of robotic trajectories in Cartesian space. They have the same dimensions and have similar shapes.

In the learning phase, the ELM is used to learn the data of robot trajectory. Half of the data is used as training data, and the other data is used as test data. After learning the neural network, we can get a new set of robot trajectory data, which is shown in Fig. 6.24. Compared with the other trajectory, the trajectory learned by ELM is more smooth.

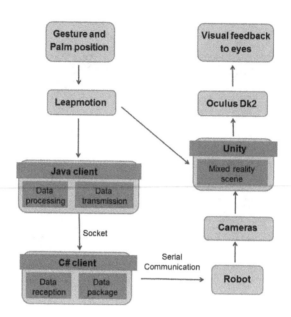

FIGURE 6.20 The system structure.

6.6.4 WRITING TASK

The Baxter robot is used to verify the effectiveness of the proposed method in writing task, whose arm has 7-DOFs. A marker pen is attached to the gripper of Baxter. The operator physically guides the Baxter to write a Chinese character in a flat paper by holding the marker pen. The experimental setup is shown in Fig. 6.25.

Regarding our experimental platform, Visual Studio 2013 and OpenCV library are used in Windows 10 operating system. The experiment environment is an adequately illuminated, indoor environment. During the teaching process, the operator demonstrates five times to writing the complete Chinese character "Mu". To end this, we have four separate single primitives, which are generalized by DMP to regroup other Chinese words. There is a self-made implementation running in a remote PC to control the recording and playback of the trajectories of Baxter by defining any text for the locally outputted trajectory files via UDP. In addition, this remote PC is installed with the software Dragon NaturallySpeaking, which transfers the voice signals to text signals, to generate robot motion control commands.

The demonstration process is repeated five times with the joint W_2 fixed, and we record the values of the joints S_0, E_1, W_0, and W_1, respectively. Then the demonstration data is used for the training of the modified DMP. The parameters of the DMP model are set as: $\tau = 1$, $K = 20$, $a = 8$. The GMM has a strong trajectory coding ability for complex tasks. This paper uses the tablet to acquire the writing data in the teaching mode, applies the GMM-based imitative learning to learn the writing skill, and obtains the generalized output through the GMR, which is performed manually

FIGURE 6.21 The operator controls the robot by LAEP Motion Controller.

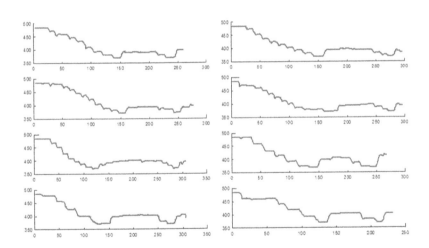

FIGURE 6.22 The Z coordinate of the robot end effector in the 8 demonstration.

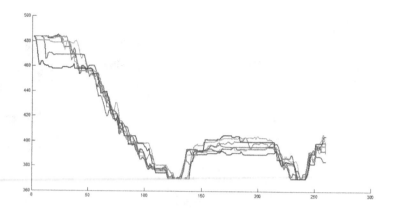

FIGURE 6.23   The data of Z coordinate aligned by DTW algorithm.

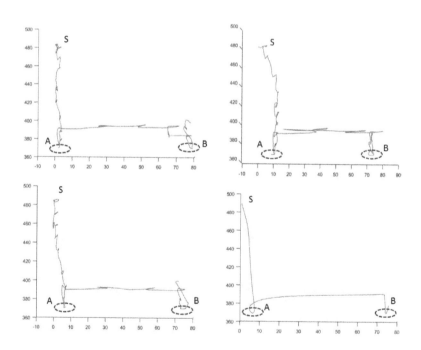

FIGURE 6.24   Three samples of the robot trajectories and the robot trajectory learned by ELM.

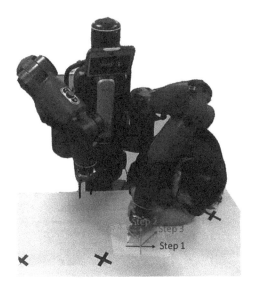

FIGURE 6.25 The experimental setup for the Chinese character writing task. Step 1: across stroke; Step 2: vertical stroke; Step 3: left-falling stroke; Step 4: right-falling stroke.

by Baxter robot. Based on the operator's demonstrated strokes, the second, third, and fourth strokes of the Chinese character "Mu" are chosen to be generalized.

The experimental trajectories are plotted using Matlab. In order to do this, the five recorded movement trajectories for each stroke are saved in Cartesian space, where we use K-means method to initial the analyzed data, and we apply the EM algorithm to obtain the GMMs. We use the DTW method to align the five recorded trajectories, here the first curve is chosen as the reference to be aligned with others.

It can be seen that the reconstructed trajectories by GMR using Matlab, and the five demonstrated trajectories in Fig. 6.26. Here we take the second step to be spatially generalized, which is the across stroke. Using the GMM-based imitative learning, the trajectory can be continuously used to write the Chinese character "Mu". The blue dotted line is the teaching trajectory by demonstration, the black solid line is the generated trajectory, and the red solid line is the generalized result after DMP and GMM coding. Next, the first stroke and all the other generalized strokes are able to form a new Chinese character "Bu" by using the verbal commands orderly as shown in Fig. 6.27. The demonstration process records the data of six joints S_0, S_1, E_0, E_1, W_0, and W_1. The joint W_2 is fixed to value 0. This dataset above is used to train the modified DMP. Fig. 6.28 illustrates the training result.

We can draw the conclusion that the maximum and the minimum values of all the joints angles between the demonstration and generalization, in some special time points, spaced about 0.04 radians apart, which means that the range of the arm movement of the robot in those two situations is differing at about 0.04, which leads to different arm motions in two different positions. The movement of joints S_0 to W_1 are regenerated through the teaching process, which enables the robot to perform the Chinese character writing task successfully, as shown in Fig. 6.27, and synthesizes the features of our proposed technology as well. As shown in Fig. 6.26, smooth

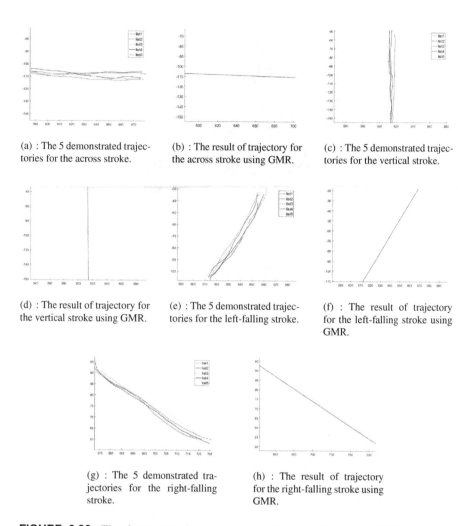

(a) : The 5 demonstrated trajectories for the across stroke.

(b) : The result of trajectory for the across stroke using GMR.

(c) : The 5 demonstrated trajectories for the vertical stroke.

(d) : The result of trajectory for the vertical stroke using GMR.

(e) : The 5 demonstrated trajectories for the left-falling stroke.

(f) : The result of trajectory for the left-falling stroke using GMR.

(g) : The 5 demonstrated trajectories for the right-falling stroke.

(h) : The result of trajectory for the right-falling stroke using GMR.

FIGURE 6.26 The demonstrated and reconstructed trajectories of the "Mu" character strokes, x axis represents x direction and y axis represents y direction.

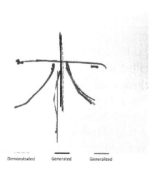

FIGURE 6.27 The initial and generalized Chinese character.

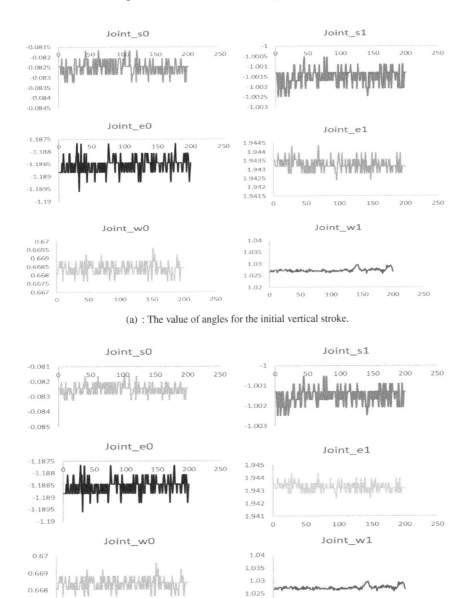

(a) : The value of angles for the initial vertical stroke.

(b) : The value of angles for the generalized vertical stroke.

FIGURE 6.28   The comparison of the joints angles for the vertical stroke of the "Mu" character with or without DMP, x axis represents time (in seconds) and y axis represents joint angles (in radians).

curves are obtained from multiple demonstrations using the modified DMP. Hence, the robot performs the writing tasks well after being taught by demonstration.

6.7 CONCLUSION

In this chapter, KF-based sensor fusion is applied to obtain improved performance. In order to do this, a Kinect sensor is utilized to capture the motion of the operator's arm with a vector approach. The vector approach can precisely calculate the angular data of human arm joints by selecting five out of seven joints on each arm. Then, the angular velocity of human operator's arm can be measured by a pair of MYO armbands worn on the operator's arm. The continuous-time KF method output the designed data with less error; after that, the data will be applied to the joints of the simulated Baxter robot for teleoperation. We also developed a teleoperation system based on visual interaction and put forward a robot teaching method based on ELM. In the teleoperation system, Kinect is used to control the robot in V-REP by the human body motion. Through learning and training, the robot can reproduce the trajectory which is provided by the RBF network. The experimental results show that a robot can learn the motion from the human demonstration in a natural manner. And this method does not require analytical modeling of a robot, so it can be customized for various types of robots.

REFERENCES

1. S Mohammad Khansari-Zadeh and Aude Billard. Learning stable nonlinear dynamical systems with Gaussian mixture models. *IEEE Trans. Robot.*, 27(5):943–957, 2011.
2. Chunxu Li, Chenguang Yang, Jian Wan, Andy SK Annamalai, and Angelo Cangelosi. Teleoperation control of Baxter robot using Kalman filter-based sensor fusion. *Systems Science & Control Engineering*, 5(1):156–167, 2017.
3. Chenguang Yang, Yiming Jiang, Zhijun Li, Wei He, and Chun-Yi Su. Neural control of bimanual robots with guaranteed global stability and motion precision. *IEEE Transactions on Industrial Informatics*, 13(3):1162–1171, 2016.
4. Julia Berg and Shuang Lu. Review of interfaces for industrial human-robot interaction. *Current Robotics Reports*, 1(2):27–34, 2020.
5. Ayan Ghosh, D Alonso Paredes Soto, Sandor M Veres, and JA Rossiter. Human robot interaction for future remote manipulations in industry 4.0. In *IFAC-PapersOnLine*. International Federation of Automatic Control (IFAC), 2020.
6. Guoqing Tang, Zidong Wang, and AL Williams. On the construction of an optimal feedback control law for the shortest path problem for the Dubins car-like robot. In *System Theory, 1998. Proceedings of the Thirtieth Southeastern Symposium on*, pages 280–284. IEEE, 1998.
7. Kerstin Dautenhahn. Socially intelligent robots: dimensions of human–robot interaction. *Philosophical Transactions of the Royal Society B: Biological Sciences*, 362(1480):679–704, 2007.
8. Parker Owan, Joseph Garbini, and Santosh Devasia. Faster confined space manufacturing teleoperation through dynamic autonomy with task dynamics imitation learning. *IEEE Robotics and Automation Letters*, 5(2):2357–2364, 2020.

9. Cynthia L Breazeal. *Sociable machines: Expressive social exchange between humans and robots*. PhD thesis, Massachusetts Institute of Technology, 2000.

10. Nicholas Paul Joseph Allec and Xiaoyi Mu. Motion and gesture input from a wearable device, April 16 2020. US Patent App. 16/706,353.

11. João André, João Lopes, Manuel Palermo, Diogo Gonçalves, Ana Matias, Fátima Pereira, José Afonso, Eurico Seabra, João Cerqueira, and Cristina Santos. Markerless gait analysis vision system for real-time gait monitoring. In *2020 IEEE International Conference on Autonomous Robot Systems and Competitions (ICARSC)*, pages 269–274. IEEE, 2020.

12. Hitesh Reddivari, Chenguang Yang, Zhaojie Ju, P Liang, Z Li, and B Xu. Teleoperation control of Baxter robot using body motion tracking. In *Multisensor Fusion and Information Integration for Intelligent Systems (MFI), 2014 International Conference on*, pages 1–6, 2014.

13. YuKang Liu, YuMing Zhang, Bo Fu, and Ruigang Yang. Predictive control for robot arm teleoperation. In *Industrial Electronics Society, IECON 2013-39th Annual Conference of the IEEE*, pages 3693–3698, 2013.

14. Inseong Jo, Younkyu Park, and Joonbum Bae. A teleoperation system with an exoskeleton interface. In *2013 IEEE/ASME International Conference on Advanced Intelligent Mechatronics*, pages 1649–1654, 2013.

15. Zhangfeng Ju, Chenguang Yang, Zhijun Li, Long Cheng, and Hongbin Ma. Teleoperation of humanoid Baxter robot using haptic feedback. In *Multisensor Fusion and Information Integration for Intelligent Systems (MFI), 2014 International Conference on*, pages 1–6. IEEE, 2014.

16. Fuwen Yang, Zidong Wang, and YS Hung. Robust Kalman filtering for discrete time-varying uncertain systems with multiplicative noises. *IEEE Transactions on Automatic Control*, 47(7):1179–1183, 2002.

17. Brenna D Argall, Sonia Chernova, Manuela Veloso, and Brett Browning. A survey of robot learning from demonstration. *Robotics and autonomous systems*, 57(5):469–483, 2009.

18. Thomas B Sheridan. Human–robot interaction: status and challenges. *Human factors*, 58(4):525–532, 2016.

19. Chenguang Yang, Peidong Liang, Zhijun Li, Arash Ajoudani, Chun-Yi Su, and Antonio Bicchi. Teaching by demonstration on dual-arm robot using variable stiffness transferring. In *2015 IEEE International Conference on Robotics and Biomimetics (ROBIO)*, pages 1202–1208. IEEE, 2015.

20. Hooman Lee, Joongbae Kim, and Taewoo Kim. A robot teaching framework for a redundant dual arm manipulator with teleoperation from exoskeleton motion data. In *2014 IEEE-RAS International Conference on Humanoid Robots*, pages 1057–1062. IEEE, 2014.

21. Ekaterina Tolstaya, Fernando Gama, James Paulos, George Pappas, Vijay Kumar, and Alejandro Ribeiro. Learning decentralized controllers for robot swarms with graph neural networks. In *Conference on Robot Learning*, pages 671–682, 2020.

22. Pedro Neto, Dário Pereira, J Norberto Pires, and A Paulo Moreira. Real-time and continuous hand gesture spotting: An approach based on artificial neural networks. In *2013 IEEE International Conference on Robotics and Automation*, pages 178–183. IEEE, 2013.

23. Klaus Neumann, Andre Lemme, and Jochen J Steil. Neural learning of stable dynamical systems based on data-driven Lyapunov candidates. In *2013 IEEE/RSJ International Conference on Intelligent Robots and Systems*, pages 1216–1222. IEEE, 2013.

24. Peidong Liang, Lianzheng Ge, Yihuan Liu, Lijun Zhao, Ruifeng Li, and Ke Wang. An augmented discrete-time approach for human-robot collaboration. *Discrete Dynamics in Nature and Society*, 2016, 2016.

25. Shunyi Yao, Ying Wen, and Yue Lu. Hog based two-directional dynamic time warping for handwritten word spotting. In *Document Analysis and Recognition (ICDAR), 2015 13th IEEE International Conference*, pages 161–165, 2015.

26. Jorge Solis, Simone Marcheschi, Antonio Frisoli, Carlo Alberto Avizzano, and Massimo Bergamasco. Reactive robot system using a haptic interface: an active interaction to transfer skills from the robot to unskilled persons. *Advanced robotics*, 21(3-4):267–291, 2007.

27. Yaodong Zhang, Kiarash Adl, and James Glass. Fast spoken query detection using lower-bound dynamic time warping on graphical processing units. In *Acoustics, Speech and Signal Processing (ICASSP), 2012 IEEE International Conference*, pages 5173–5176, 2012.

28. James Altman. Taming the dragon effective use of dragon naturallyspeaking speech recognition software as an avenue to universal access. *Writing & Pedagogy*, 5(2):333–348, 2014.

7 Neural Network-Enhanced Robot Manipulator Control

7.1 INTRODUCTION

The controller design is important in the process of robot skill reproduction [17]. Admittance control has been widely used in human-robot interaction, which can generate robot motions based on the human force [8, 9, 18]. Thus, we can use the admittance control to achieve human-guided teaching. The admittance control [19] exploits the end-effector position controller to track the output of an admittance model. Most studies on admittance control have not considered the human factor, which is an important part of this control loop. The interaction force between the robot and the human can be used to recognize the human intention, and to improve further the interaction safety and user experience [10]. The human force has been employed to compute the desired movement trajectory of the human, which is used to get the performance index of the admittance model [1]. The unknown human dynamics has been considered in the control loop. The human transfer function and the admittance model were formulated as a Wiener filter [20], and a task model was used to estimate the human intent. The estimate law based on Kalman filter was used to tune the parameters of the admittance model [21]. However, the task model was assumed as a certain linear system, which is unreasonable because the estimated human motions should be different for each individual due to different motion habits. Thus, a task model that involves the human characteristics needs to be developed.

The DMP [22, 3] offers a compact implementation of the motion model using DS, enabling the robot to reproduce human-like motions. The traditional DMP model can only be used to handle a single demonstration. However, multiple demonstrations are necessary because optimal motion is difficult to obtain through only one-time teaching, even for an expert [4]. To recognize more features, the data captured from multiple demonstrations should be integrated into the nonlinear term of the DMP model. The probabilistic methods have shown their feasibility to tackle this problem [2, 23, 5]. For example, the GMR [24], which is based on a probabilistic model named the GMM, has been employed to extract the important features of the task. The GMR has also been utilized to construct the DS model called the stable estimator of DSs for stable motions. Inspired by this work and further considering the fitting performance of the GMM, the fuzzy GMM (FGMM) [6] is employed to fuse the features of multiple demonstrations into the nonlinear term of the DMP, which has been proposed to improve the learning efficiency of the active curve axis GMM and has shown better nonlinearity fitting performance than the conventional GMM [7].

A novel regression algorithm for the FGMM is further developed to retrieve the nonlinear term, according to the geometric significance of the GMR. In addition, only reproducing and generalizing the motion profile of demonstrations could not meet the need of completing complex tasks in various working environments. Therefore, the skill representation can be extended by including force features modeled by DMPs.

In practice, only considering the motion modeling is insufficient for a stable LfD framework because of the dynamic and unstructured environments, which will result in many disturbances and variation of the robot's dynamics. If the manipulator is controlled using a model-based control method, the situation mentioned above will affect the control performance and even make the system unstable [11]. Considering the uncertainties of the robot dynamics, various approximation tools [25, 26], such as NNs [27] and fuzzy logic systems [12] have been integrated into the control design to approximate the uncertainties. Recently, NNs has served as a promising computational tool in various fields; for example, the primal-dual neural network has been employed to solve a complicated quadratic programming problem [13].

For the dynamics controllers that employ the NNs, the learning efficiency is an important aspect that should be considered, because there is a tradeoff between the approximation accuracy and the efficiency of the NNs. The backpropagation NN (BPNN) has been utilized to approximate the unknown nonlinear function in the model of the vibration suppression device [28]. The radial basis function NNs (RBFNNs) was utilized to approximate the unknown nonlinearity of the telerobot system [16]. Compared to BPNN, the learning procedure of RBFNN is based on local approximation; thus, RBFNN can avoid getting stuck in the local optimum and has a faster convergence rate. Besides, the number of hidden layer units of RBFNN can be adaptively adjusted during the training phase, making NNs more flexible and adaptive. Therefore, RBFNN is more appropriate for the design of real-time control. The cerebellar model articulation controller (CMAC) is a type of NNs that has been adopted widely in dynamics control design [14, 29]. The structure of the CMAC is inspired by the information processing mode of the cerebellum. This NNs is not fully connected to associative memory; thus, local weights are updated during each learning cycle to provide faster learning compared to fully connected NNs, without function approximation loss. We have also developed a CMAC-NN-based controller which can achieve accurate and stable motion under the output constraints. This constraint exists commonly in real-world robotic systems, such as nonholonomic mobile robots [30], and its effect can be compensated with the help of a barrier Lyapunov function (BLF). The CMAC is employed to approximate the unknown dynamics of the robot.

In this chapter, we will introduce a LfD approach which is enhanced by GMM and FGMM. Then, the neural networks control scheme, adaptive admittance controller, and the hybrid adaptive controller will be described. Finally, several experiments have been performed using a Baxter robot, and the results will be shown to validate of the proposed methods.

7.2   PROBLEM DESCRIPTION

The LfD framework that considers the teaching phase, the learning phase, and the reproduction phase is developed, as shown in Fig. 7.1. In the teaching phase, the adaptive admittance controller is employed so that the human tutor can smoothly guide the robot to accomplish the demonstration. In the learning phase, the DMP-based methods are used to model the robotic motion. The learned model can generalize the motion to adapt to different situations. In the reproduction phase, the NN-based trajectory tracking controller and the hybrid force-motion controller are developed to achieve accurate motion reproduction.

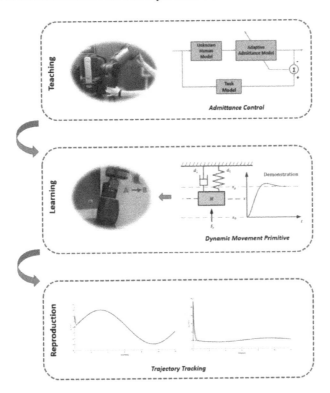

FIGURE 7.1 Overview of the LfD framework.

7.3 LEARNING FROM MULTIPLE DEMONSTRATIONS

As introduced in chapter 4, DMP is widely used to encode movements. The conventional method used for learning in a DMP model is to solve a linear regression problem, where the demonstration is assumed as the data generated from the model, and the expected nonlinear function of $f(s)$ is defined as follows

$$f^*(s) = \frac{\tau_s \ddot{x}(\nabla_s) + d_2 \dot{x}(\nabla_s)}{d_1} - (x_g - x(\nabla_s)) + (x_g - x_0)s \qquad (7.1)$$

where $x(\cdot)$ is the function of a given demonstration trajectory, ∇_s denotes the inverse function of $s(t) = s_0 \exp(-\alpha_s t / \tau_s)$.

7.3.1 GAUSSIAN MIXTURE MODEL

When n_d demonstration trajectories $\{x_i(t)\}$ are given, multiple expected nonlinear functions, $\{f_i^*(s)\}$, for $i = 1, 2, ..., n_d$, can be obtained. Then the GMR, can be employed to fuse the data obtained from these functions.

Assume that $O_b = \{o_1, ..., o_t, ..., o_{np}\}$ with $o_t = [o_{1t}, o_{2t}]^T \in R^2$ is an observed dataset generated from the mapping sets $\{f_1^*, ..., f_n^*\}$ through discretization, where $o_{1t} \in s, o_{2t} \in f_i^*(o_{1t})$, and n_p is the number of the data o_t. The distribution of O_b is modeled by the GMM with finite Gaussian distributions, the probability density of which is [5]

$$p(O_b|\Theta) = \prod_{t=1}^{n_p} p(o_t|\Theta) = \prod_{t=1}^{n_p} (\sum_{i=1}^{n_g} \alpha_i p(o_t|\theta_i)) \tag{7.2}$$

where $\Theta = (\alpha_1, ..., \alpha_{ng}, \theta_1, ..., \theta_{ng})$, $\alpha_i \in R$ is the mixing weight with $\sum_{i=1}^{ng} \alpha_i = 1$, n_g is the number of the Gaussian distributions, and $\theta_i = (\mu_i, \sigma_i)$ is the parameter of the i-th Gaussian distribution

$$p(O_b|\theta_i) = \frac{\exp(-0.5(o_t - \mu_i)^T \sigma_i^{-1}(o_t - \mu_i))}{2\pi \sqrt{|\alpha_i|}} \tag{7.3}$$

where $\mu_i \in R^2$ is the mean and $\sigma_i \in R^{2 \times 2}$ is the co-variance matrix

$$\mu_i = \begin{bmatrix} \mu_{1i} \\ \mu_{2i} \end{bmatrix}, \quad \sigma_i = \begin{bmatrix} \sigma_{1i} & \sigma_{12i} \\ \sigma_{12i} & \sigma_{2i} \end{bmatrix} \tag{7.4}$$

The maximum likelihood estimation is employed to estimate the parameters of the GMM. Then the GMR is utilized to retrieve the composite expected function, which is defined as

$$\bar{f}^*(s) = \sum_{i=1}^{n_g} \beta_i(s)\eta_i(s) \tag{7.5}$$

with

$$\beta_i(s) = \frac{\alpha_i \mathscr{G}(s|\mu_{1i}, \sigma_{1i})}{\sum_{i=1}^{n_g} \alpha_i \mathscr{G}(s|\mu_{1i}, \sigma_{1i})} \tag{7.6}$$

$$\eta_i(s) = \mu_{2i} + \frac{\sigma_{12i}}{\sigma_{1i}}(s - \mu_{1i}) \tag{7.7}$$

where $\mathscr{G}(s|\mu_{1i}, \sigma_{1i})$ denotes the 1-D Gaussian distribution function with the mean μ_{1i} and the variance σ_{1i}. Using the data obtained from (7.10) and the LSM, the weights can be estimated.

7.3.2 FUZZY GAUSSIAN MIXTURE MODEL

Considering the nonlinearity of the demonstrations, the conventional GMM is replaced by the FGMM to improve the fitting performance, which is based on the generalized single Gaussian model (SGM).

1) Generalized SGM: For a 2-D generalized SGM, one of its axes is beeline, and the other is bent, which correspond to the conventional Gaussian model and the AcaG model, respectively. The 2-D plane that the curve principal axis is located in is referred to as the principal plane. The observations are first transformed to the principal plane by the principal component analysis (PCA) method

$$r_t = Q(o_t - T) \tag{7.8}$$

where the PCA is used for coordinate transformation, $T \in R^2$ is the translation vector that includes the means of the sample, $Q \in R^{2 \times 2}$ is the rotation matrix which is composed of the eigenvectors of the covariance matrix, and $r_t \in R^2$ is the transformed point of o_t which is located in the principal plane. The curve principal axis is chosen as a parabola to fit the point set $\{r_t\}$

$$\bar{r}_{2t} = a_p \bar{r}_{1t}^2 + b_p \tag{7.9}$$

where $\bar{r}_t = [\bar{r}_{1t}, \bar{r}_{2t}]$ denotes the point in the curve principal axis, and $a_p, b_p \in R$ are computed by using the weighted least squares method.

To derive the probability density of O_b, we first consider the AcaG model, the axis of which is located in the curve. Assume that the projection points of r_t in the curve is $z_t = \{z_{1t}, ..., z_{jt}, ..., z_{J_t t}\}$, where J_t is the number of the projection points of r_t. In the principal plane, the center of the AcaG is $(0,0)$; thus, the probability of the point z_{jt} is [7]

$$p_1(z_{jt}) = \frac{\exp\left(-0.5 l_{aj}^2(r_t)\bar{\sigma}_1^{-1}\right)}{\sqrt{2\pi|\bar{\sigma}_1|}} \tag{7.10}$$

where $\bar{\sigma}_1 \in R$ is the variance of the AcaG model, and $l_{aj}(r_t)$ denotes the arc length between $(0,0)$ and z_{jt}.

For the conventional Gaussian model in generalized SGM, its center is located in projection point z_{jt}. Therefore, given z_{jt}, the probability density is [7]

$$p_2(o_t|z_{jt}) = \frac{\exp\left(-0.5 l_{bj}^2(r_t)\bar{\sigma}_2^{-1}\right)}{\sqrt{2\pi|\bar{\sigma}_2|}} \tag{7.11}$$

where $\bar{\sigma}_2 \in R$ is the variance of the conventional Gaussian model, and $l_{bj}(r_t)$ denotes the distance between z_{jt} and r_t. Then the probability distribution of the generalized SGM can be computed by Ref. [7]

$$p(o_t|\theta) = \sum_{j=1}^{J_t} \frac{\exp\left(-0.5 l_{aj}^2(r_t)\bar{\sigma}_1^{-1} - 0.5 l_{bj}^2(r_t)\bar{\sigma}_2^{-1}\right)}{2\pi\sqrt{|\bar{\sigma}_1\bar{\sigma}_2|}} \tag{7.12}$$

where $\theta = (Q, T, a_p, b_p, \bar{\sigma}_1, \bar{\sigma}_2)$ includes the parameters of the generalized SGM.

2) Fuzzy GMM: The FGMM constructs a novel mixture model by replacing the original GMM with SGM, and introduces the fuzzy membership in the EM (Expectation Maximization) algorithm to effectively solve the problem of the parameter

estimation [6]. The iterative procedure of the EM algorithm for FGMM improves the learning performance of the conventional method.

Since the GMR employed for the learning of the DMP is based on the conventional GMM, a new regression algorithm for FGMM should be derived. The parameters of the FGMM (T_i, Q_i) involve the means of the Gaussian models in the original data space, i.e., the information of the means, which is employed in GMR, is implied in (T_i, Q_i) through the PCA transformation. Therefore, the regression algorithm for FGMM cannot be derived directly.

To derive the regression algorithm for FGMM, the geometrical significance of GMR is first discussed. The result of the GMR can be rewritten as

$$\bar{f}^*(s) = \sum_{i=1}^{n_g} \beta_i(s) (a_{ri}s + b_{ri}) \tag{7.13}$$

where $a_{ri} = \sigma_{12i}/\sigma_{1i}$ and $b_{ri} = \mu_{2i} - \mu_{1i}\sigma_{12i}/\sigma_{1i}$. Note that the item $(a_{ri}s + b_{ri})$ is a linear function, where the axis of the i-th Gaussian model is located. Therefore, the GMR can be regarded as the weighted summation of a set of linear functions, where the weight $\beta_i(s)$ is the normalized probability of the Gaussian model along the axis of the input.

For the FGMM, the corresponding axis, denoted by $y_{ci}(s)$, can be obtained through the PCA inverse transformation of the curve principal axis

$$\bar{c}_{it} = Q_i^{-1}c_{it} + T_i \tag{7.14}$$

where $c_{it} \in R^2$ and $\bar{c}_{it} \in R^2$ denote the points in the curve principal axis and in axis $y_{ci}(s)$, respectively. The weight of point $\bar{c}_{it}$ is computed by

$$\beta_{ci}(s) = \frac{\alpha_i \sum_{j=1}^{J_{it}} \mathscr{G}\left(l_{ij(1)}(c_{it})|0, \bar{\sigma}_{1i}\right)}{\sum_{i=1}^{n_g} \alpha_i \sum_{j=1}^{J_{it}} \mathscr{G}\left(l_{ij(1)}(c_{it})|0, \bar{\sigma}_{1i}\right)} \tag{7.15}$$

According to the geometrical significance discussed above, the regression for FGMM can be written as

$$\bar{F}_R(s) = \sum_{i=1}^{n_g} \beta_{ci}(s)y_{ci}(s) \tag{7.16}$$

7.4 NEURAL NETWORKS TECHNIQUES

7.4.1 RADIAL BASIS FUNCTION NEURAL NETWORK

RBFNN is an effective tool to approximate any continuous function $g : R^m \to R$ as follows [27]:

$$g(\vartheta) = W^T S(\vartheta) + \varepsilon(\vartheta), \quad \forall \vartheta \in \Omega_\vartheta \tag{7.17}$$

where $\vartheta \in \Omega_\vartheta \subset R^m$ denotes the input vector, $W = [\omega_1, \omega_2, ..., \omega_N]^T \in R^N$ is the ideal NN weight vector and N is the number of NN nodes. The approximation error $\varepsilon(\vartheta)$

is bounded. $S(\vartheta) = [s_1(\vartheta), s_2(\vartheta), ..., s_N(\vartheta)]^T$ is a nonlinear vector function, where $s_i(\vartheta)$ is defined as a Gaussian function:

$$s_i(\vartheta) = \exp\left[-\frac{(\vartheta - \kappa_i)^T(\vartheta - \kappa_i)}{\chi_i^2}\right], i = 1, 2, ..., N \qquad (7.18)$$

where $\kappa_i = [\kappa_{i1}, \kappa_{i2}, ..., \kappa_{im}]^T \in R^m$ represents the center of the Gaussian function and χ_i^2 is the variance. The ideal weight vector W is defined as follows:

$$W = \arg\min_{\hat{W} \in R^N} \left\{ \sup_{\vartheta \in \Omega_\vartheta} \left| g(\vartheta) - \hat{W}^T S(\vartheta) \right| \right\} \qquad (7.19)$$

which minimizes the approximation error for all $\vartheta \in \Omega_\vartheta$.

7.4.2 CEREBELLAR MODEL ARTICULATION NEURAL NETWORKS

The CMAC-NN is an efficient functional approximator with fast learning rate. The structure of the NN consists of five spaces. Considering the estimation error, any C1-function $H(X) : R^{n_I} \to R^{n_o}$ approximated by the CMAC-NN is presented as follows [15]

$$H(X) = W^T B(X) + \varepsilon \qquad (7.20)$$

where $X = [X_1, ..., X_{n_I}]^T \in R^{n_I}$ is the input vector, $W \in R^{n_I \times n_o}$ is the weight matrix, n_I is the number of the layouts of the NN, $B(X) = [B_1(X), ..., B_{n_I}(X)]^T \in R^{n_I}$ is the receptive field function vector with

$$B_k(X) = \exp\left[\sum_{i=1}^{n_I} \frac{-(X_i - \bar{b}_{ik})^2}{2\bar{a}_{ik}^2}\right] \qquad (7.21)$$

where $\bar{b}_{ik}$ is the mean and $\bar{a}_{ik}^2$ is the variance. ε is the approximation error with a known bound.

7.5 ROBOT MANIPULATOR CONTROLLER DESIGN

7.5.1 NN-BASED CONTROLLER FOR ROBOTIC MANIPULATOR

The dynamics of a robot manipulator with n-link is described as follows:

$$M(\theta)\ddot{\theta} + C(\theta, \dot{\theta})\dot{\theta} + G_0(\theta) + \tau_p = \tau \qquad (7.22)$$

where $\theta \in R^n$, $\dot{\theta} \in R^n$, and $\ddot{\theta} \in R^n$ are the joint position, joint velocity, and joint acceleration, respectively. τ is the control torque and τ_p is the torque caused by the payload. $M(\theta) \in R^{n \times n}$ denotes the inertia matrix, which is symmetric positive-definite. $C(\theta, \dot{\theta}) \in R^{n \times n}$ represents the Coriolis and centripetal torque matrix and $G_0(\theta) \in R^n$ is the gravity vector. The matrices $M(\theta)$ and $C(\theta, \dot{\theta})$ satisfy

$$s^T(\dot{M} - 2C)s = 0, \ \forall s \in R^n \qquad (7.23)$$

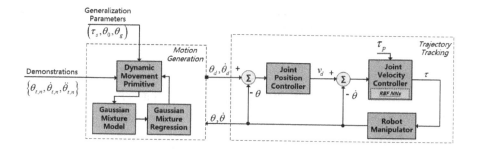

FIGURE 7.2 Block diagram of the proposed system.

The controller design includes the design of the joint position controller and the joint velocity controller, as shown in Fig. 7.2. RBFNN is used in the latter to approximate the uncertain dynamics.

1) Joint position controller: The joint position tracking error is defined as $e_p = [e_{p1}, e_{p2}, ..., e_{pn}]^T = \theta - \theta_d$, where $\theta_d = [\theta_{d1}, \theta_{d2}, ..., \theta_{dn}]^T \in R^n$ is the reference trajectory, which is smooth and bounded. The error transformation function is introduced as follows:

$$e_{pi}(t) = \delta(t) H_i \left(L_i \left(\frac{e_{pi}(t)}{\delta(t)} \right) \right), \quad i = 1, 2, ..., n \tag{7.24}$$

where $\delta(t) = (\delta_0 - \delta_\infty)e^{-at} + \delta_\infty$ represents the tracking performance requirement, and $H_i(z)$ is defined as

$$H_i(z) = \begin{cases} \dfrac{e^z - \sigma}{1 + e^z}, & e_{pi}(0) \geq 0 \\[2mm] \dfrac{\sigma e^z - 1}{1 + e^z}, & e_{pi}(0) < 0 \end{cases} \tag{7.25}$$

and $L_i(z)$ is the inverse function of $H_i(z)$:

$$L_i(z) = \begin{cases} \ln \dfrac{z + \sigma}{1 - z}, & e_{pi}(0) \geq 0 \\[2mm] \ln \dfrac{z + 1}{\sigma - z}, & e_{pi}(0) < 0 \end{cases} \tag{7.26}$$

The parameters δ_0, δ_∞, a, and σ are positive constants, which are used to adjust the control performance, and $\delta_\infty < \delta_0$. The joint position controller is used to generate the desired joint velocity, which is designed as [16].

$$v_{di} = -k_1 \delta(t) \phi_i(t) + \dot{\theta}_{di}(t) + \frac{\dot{\delta}(t)}{\delta(t)} e_{pi}(t) \tag{7.27}$$

where k_1 is a positive constant, and

$$\phi_i(t) = L_i \left(\frac{e_{pi}(t)}{\delta(t)} \right) \tag{7.28}$$

If $\phi_i(t)$ is bounded, then the following is obtained:

$$\begin{cases} -\sigma\delta(t) < e_{pi}(t) < \delta(t), & e_{pi}(0) > 0 \\ -\delta(t) < e_{pi}(t) < \sigma\delta(t), & e_{pi}(0) < 0 \end{cases} \tag{7.29}$$

Therefore, the function $\delta(t)$ determines the boundary of error $e_{pi}(t)$ and the transient performance of the controller.

2) *Joint velocity controller*: The joint velocity controller aims to generate the control torque to track the desired joint velocity $v_d = [v_{d1}, v_{d2}, ..., v_{dn}]^T$. Let us define the joint velocity error as $e_v = \dot{\theta} - v_d$ and $G(\theta) = G_0(\theta) + \tau_p$. Then, the control torque is designed as follows[16]:

$$\tau = -k_2 e_v - \frac{\dot{Q}(\phi(t))\phi(t)}{\delta(t)} + \hat{M}\dot{v}_d + \hat{C}v_d + \hat{G} + \hat{r} \tag{7.30}$$

where k_2 is a positive constant, and

$$\begin{aligned} \dot{Q}(\phi(t)) &= diag(\dot{L}_1(H_1(\phi_1(t))), ..., \dot{L}_n(H_n(\phi_n(t)))) \\ \phi(t) &= [\phi_1(t), \phi_2(t), ..., \phi_n(t)]^T \end{aligned} \tag{7.31}$$

The matrices $\hat{M}(\theta)$, $\hat{C}(\theta, \dot{\theta})$, $\hat{G}(\theta)$, and $\hat{r}(\theta, \dot{\theta}, v_d, \dot{v}_d)$ are the estimates of $M(\theta)$, $C(\theta, \dot{\theta})$, $G(\theta)$, and $r(\theta, \dot{\theta}, v_d, \dot{v}_d)$, respectively, where $r(\theta, \dot{\theta}, v_d, \dot{v}_d)$ is defined later in (7.40).

Substituting (7.36) into (7.28), we can obtain the closed-loop dynamics equation:

$$M\dot{e}_v + Ce_v + k_2 e_v + \frac{\dot{Q}(\phi(t))\phi(t)}{\delta(t)} - \hat{r} = $$
$$- (M - \hat{M})\dot{v}_d - (C - \hat{C})v_d - (G - \hat{G}) \tag{7.32}$$

Then, RBFNN is utilized to approximate $M(\theta), C(\theta, \dot{\theta}), G(\theta)$, and $r(\theta, \dot{\theta}, v_d, \dot{v}_d)$:

$$\begin{aligned} M(\theta) &= W_M^T S_M(\theta) + \varepsilon_M \\ C(\theta, \dot{\theta}) &= W_C^T S_C(\theta, \dot{\theta}) + \varepsilon_C \\ G(\theta) &= W_G^T S_G(\theta) + \varepsilon_G \\ r(\theta, \dot{\theta}, v_d, \dot{v}_d) &= W_r^T S_r(\theta, \dot{\theta}, v_d, \dot{v}_d) + \varepsilon_r \end{aligned} \tag{7.33}$$

where $W_M \in R^{nN \times n}$, $W_C \in R^{2nN \times n}$, $W_G \in R^{nN \times n}$, and $W_r \in R^{4nN \times n}$ are the ideal NN weight matrices. $S_M(\theta) \in R^{nN \times n}$, $S_C(\theta, \dot{\theta}) \in R^{2nN \times n}$, $S_G(\theta) \in R^{nN \times n}$, and $S_r(\theta, \dot{\theta}, v_d, \dot{v}_d) \in R^{4nN \times n}$ are the RBF matrices. ε_M, ε_C, ε_G, and ε_r are the approximation errors. The function $r(\theta, \dot{\theta}, v_d, \dot{v}_d)$ is defined as

$$r(\theta, \dot{\theta}, v_d, \dot{v}_d) = \varepsilon_M \dot{v}_d + \varepsilon_C v_d + \varepsilon_G \tag{7.34}$$

The estimates of M, C, G, and r are written as follows:

$$\begin{aligned} \hat{M}(\theta) &= \hat{W}_M^T S_M(\theta) \\ \hat{C}(\theta, \dot{\theta}) &= \hat{W}_C^T S_C(\theta, \dot{\theta}) \\ \hat{G}(\theta) &= \hat{W}_G^T S_G(\theta) \\ \hat{r}(\theta, \dot{\theta}, v_d, \dot{v}_d) &= \hat{W}_r^T S_r(\theta, \dot{\theta}, v_d, \dot{v}_d) \end{aligned} \tag{7.35}$$

Substituting (7.41) into (7.38) and defining $\tilde{W}_{(\cdot)} = W_{(\cdot)} - \hat{W}_{(\cdot)}$, we have

$$
\begin{aligned}
M\dot{e}_v + Ce_v + k_2 e_v + \frac{\dot{Q}(\phi(t))\phi(t)}{\delta(t)} = \\
-\tilde{W}_M^T S_M \dot{v}_d - \tilde{W}_C^T S_C v_d - \tilde{W}_G^T S_G - \tilde{W}_r^T S_r - \varepsilon_r
\end{aligned}
\tag{7.36}
$$

3) Stability analysis: Consider a Lyapunov function as follows:

$$
V = V_1 + V_2
\tag{7.37}
$$

with

$$
V_1 = \frac{1}{2}\phi^T(t)\phi(t)
\tag{7.38}
$$

and

$$
\begin{aligned}
V_2 = & \frac{1}{2}e_v^T M e_v + \frac{1}{2}\mathrm{tr}(\tilde{W}_M^T \Gamma_M^{-1}\tilde{W}_M + \tilde{W}_C^T \Gamma_C^{-1}\tilde{W}_C) \\
& + \frac{1}{2}\mathrm{tr}(\tilde{W}_G^T \Gamma_G^{-1}\tilde{W}_G + \tilde{W}_r^T \Gamma_r^{-1}\tilde{W}_r)
\end{aligned}
\tag{7.39}
$$

where Γ_M^{-1}, Γ_C^{-1}, Γ_G^{-1}, and Γ_r^{-1} are positive definite matrices.
Taking the derivatives of V_1 and V_2, we have

$$
\dot{V}_1 = \frac{\phi^T(t)\dot{Q}(\phi(t))e_v(t)}{\delta(t)} - k_1\phi^T(t)\dot{Q}(\phi(t))\phi(t)
\tag{7.40}
$$

and

$$
\begin{aligned}
\dot{V}_2 = & -e_v^T k_2 e_v - e_v^T \varepsilon_r - \frac{\phi^T(t)\dot{Q}(\phi(t))e_v(t)}{\delta(t)} \\
& - \mathrm{tr}\left[\tilde{W}_M^T(S_M \dot{v}_d e_v^T + \Gamma_M^{-1}\dot{\hat{W}}_M)\right] \\
& - \mathrm{tr}\left[\tilde{W}_C^T(S_C v_d e_v^T + \Gamma_C^{-1}\dot{\hat{W}}_C)\right] \\
& - \mathrm{tr}\left[\tilde{W}_G^T(S_G e_v^T + \Gamma_G^{-1}\dot{\hat{W}}_G)\right] \\
& - \mathrm{tr}\left[\tilde{W}_r^T(S_r e_v^T + \Gamma_r^{-1}\dot{\hat{W}}_r)\right]
\end{aligned}
\tag{7.41}
$$

Let us design the update law of the NN weights as follows:

$$
\begin{aligned}
\dot{\hat{W}}_M &= -\Gamma_M(S_M \dot{v}_d e_v^T + \rho_M \hat{W}_M) \\
\dot{\hat{W}}_C &= -\Gamma_C(S_C v_d e_v^T + \rho_C \hat{W}_C) \\
\dot{\hat{W}}_G &= -\Gamma_G(S_G e_v^T + \rho_G \hat{W}_G) \\
\dot{\hat{W}}_r &= -\Gamma_r(S_r e_v^T + \rho_r \hat{W}_r)
\end{aligned}
\tag{7.42}
$$

where ρ_M, ρ_C, ρ_G, and ρ_r are positive constants. Then, the derivative of V is written as follows[16]:

$$
\begin{aligned}
\dot{V} = &-e_v^T k_2 e_v - e_v^T \varepsilon_r - k_1 \phi^T(t) \dot{Q}(\phi(t)) \phi(t) \\
&+ \text{tr}[\rho_M \tilde{W}_M^T \hat{W}_M] + \text{tr}[\rho_C \tilde{W}_C^T \hat{W}_C] \\
&+ \text{tr}[\rho_G \tilde{W}_G^T \hat{W}_G] + \text{tr}[\rho_r \tilde{W}_r^T \hat{W}_r]
\end{aligned}
\tag{7.43}
$$

We can obtain the following inequality according to the definition of $\dot{Q}(\phi(t))$:

$$
\phi^T(t) \dot{Q}(\phi(t)) \phi(t) \geq \frac{2}{(1+\sigma)} \|\phi(t)\|^2
\tag{7.44}
$$

Using the Young's inequality, we have

$$
\begin{aligned}
\dot{V} \leq &-\frac{2k_1}{(1+\sigma)} \|\phi(t)\|^2 - (k_2 - \frac{1}{2}) \|e_v\|^2 + \varpi \\
&-\frac{\rho_M}{2} \|\tilde{W}_M\|_F^2 - \frac{\rho_C}{2} \|\tilde{W}_C\|_F^2 \\
&-\frac{\rho_G}{2} \|\tilde{W}_G\|_F^2 - \frac{\rho_r}{2} \|\tilde{W}_r\|_F^2
\end{aligned}
\tag{7.45}
$$

with

$$
\begin{aligned}
\varpi = &\frac{\rho_M}{2} \|W_M\|_F^2 + \frac{\rho_C}{2} \|W_C\|_F^2 + \frac{\rho_G}{2} \|W_G\|_F^2 \\
&+ \frac{\rho_r}{2} \|W_r\|_F^2 + \frac{1}{2} \kappa_r^2
\end{aligned}
\tag{7.46}
$$

where κ_r is the upper bound of $\|\varepsilon_r\|$ over Ω.

We have $\dot{V} \leq 0$ if $\tilde{W}_M$, $\tilde{W}_C$, $\tilde{W}_G$, $\tilde{W}_r$, $\phi(t)$, and e_v satisfy the inequality as follows:

$$
\begin{aligned}
\rho = &\frac{2k_1}{(1+\sigma)} \|\phi(t)\|^2 + (k_2 - \frac{1}{2}) \|e_v\|^2 + \frac{\rho_M}{2} \|\tilde{W}_M\|_F^2 \\
&+ \frac{\rho_C}{2} \|\tilde{W}_C\|_F^2 + \frac{\rho_G}{2} \|\tilde{W}_G\|_F^2 + \frac{\rho_r}{2} \|\tilde{W}_r\|_F^2 \geq \varpi
\end{aligned}
\tag{7.47}
$$

According to LaSalle's theorem, all closed-loop signals of the dynamics system composed of (7.25), (7.36), and (7.48) are semi-global uniformly bounded if the input signals θ_d and $\dot{\theta}_d$ are bounded. Besides, $\phi(t)$ and e_v will converge to an invariant set $\Omega_i \subseteq \Omega$:

$$
\begin{aligned}
\Omega_i = \{ (\|\phi(t)\|, \|e_v\|, \|W_M\|, \|W_C\|, \\
\|W_G\|, \|W_r\|) | \rho/\varpi \leq 1 \}
\end{aligned}
\tag{7.48}
$$

Since the signal $\phi(t)$ is bounded, the transient performance and the stability of the controller are guaranteed.

7.5.2 ADAPTIVE ADMITTANCE CONTROLLER

The prescribed admittance model is defined as follows [9]:

$$M_m \ddot{x}_m + D_m \dot{x}_m + K_m x_m = f_h \tag{7.49}$$

where M_m is a prescribed mass matrix, D_m is a prescribed damping matrix, K_m is a prescribed spring constant matrix, and f_h is the human input force. The function of this admittance model is to generate the desired robot response $x_m(t)$, which serves as the human demonstration later.

The parameters of the admittance model can be adapted based on the task model output $x_d(t)$ and the human input force $f_h(t)$ by employing the Recursive least squares (RLS) algorithm. First, the admittance model should be discretized. Define $x_m(k)$ as the value of x_m at time step k, and T_s as the sampling period. The velocity and the acceleration of $x_m(t)$ are defined as:

$$\dot{x}_m(k) = \frac{x_m(k) - x_m(k-1)}{T_s} \tag{7.50}$$

$$\ddot{x}_m(k) = \frac{x_m(k) - 2x_m(k-1) + x_m(k-2)}{T_s^2} \tag{7.51}$$

The spring constant matrix K_m is assumed as zero because the spring term will pull the motion back to the origin when the human input force is equal to zero. Then the discrete form of the admittance model is:

$$h_0 x_m(k) + h_1 x_m(k-1) + h_2 x_m(k-2) = f_h \tag{7.52}$$

where

$$h_0 = \frac{M_m}{T_s^2} + \frac{D_m}{T_s} \tag{7.53}$$

$$h_1 = -\left(\frac{2M_m}{T_s} + \frac{D_m}{T_s} \right) \tag{7.54}$$

$$h_2 = \frac{M_m}{T_s^2} \tag{7.55}$$

To employ the RLS algorithm, the model is rewritten as:

$$x_m(k) = H(k)^T Z \tag{7.56}$$

where $H = [-h_0^{-1} h_1, -h_0^{-1} h_2, h_0^{-1}]^T$. Define $\hat{H}(k)$ as the estimate of H. The RLS algorithm is applied to minimize the error between the admittance model output and the task model output, which is referred to as the reference task trajectory

$$J = \sum_k \left\| \hat{H}(k)^T Z(k) - x_d(k) \right\|^2 \tag{7.57}$$

This error specified is taken as the performance index in model estimation. The estimated update equations of RLS are defined as follows

$$R(k) = I + Z^T(k) P(k-1) Z(k) \tag{7.58}$$

$$K(k) = P(k-1)Z(k)R(k)^{-1} \tag{7.59}$$

$$\dot{P}(k) = -K(k)Z^T(k)P(k-1) \tag{7.60}$$

$$\dot{\hat{H}}(k) = K(k)\left[x_d(k) - \hat{H}(k-1)^T Z(k)\right] \tag{7.61}$$

where $R(k)$ is an auxiliary variable, I is an identity matrix, $K(k)$ is the gain, $\hat{H}(k)$ is the estimated admittance model parameter matrix, respectively. This algorithm is initialized by setting $\hat{H}(0) = 0$, and a threshold for the error is defined to guarantee the convergence of the algorithm.

7.6 EXPERIMENTAL STUDY

7.6.1 TEST OF THE ADAPTIVE ADMITTANCE CONTROLLER

In this group of experiments, the performance of adaptive admittance control is tested. An ATI force sensor is attached to the end of the left arm to detect the human force.

The push-and-pull experiment is conducted to test the performance of the adaptive admittance controller. The human tutor holds the left end-effector of the robot and moves it between two specified points along the y-axis back and forth. Two specified points are set as $y_1 = 0$ (m) and $y_2 = 0.33$ (m). The period is set as 8s. First, an admittance controller with fixed parameters is tested. The parameters is set as: $M_m = 10, D_m = 40, K_m = 0$. The human force in this process is recorded, as shown in Fig. 7.5(a). Then this group of parameters is used to initialize the parameters in adaptive admittance controller with a sampling period $T_s = 0.02$s, which are transformed in RLS algorithm as: $H_1 = 1.92, H_2 = -0.92, H_3 = 0.0004$.

We conduct comparative experiments to verify the performance of the proposed task model. In the first experiment, a linear task model is employed, the output of which is exponential type function. In the second experiment, the human tutor first demonstrates this point-to-point motions four times, and the demonstration is recorded to learn a task model using GMR. The number of the Gaussian components is set as 16. The learning result is shown in Figs. 7.3. Then this task model is used in the adaptive admittance controller. The output of the admittance model and the human force in both experiments are recorded, as shown in Figs. 7.4 and 7.5.

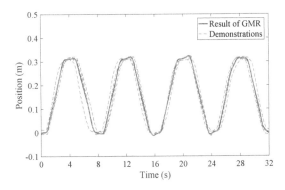

FIGURE 7.3 Learned task model using GMR.

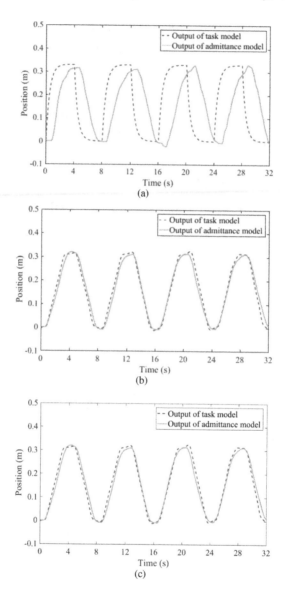

FIGURE 7.4 Model output: (a) (b) Adaptive admittance controller with linear task model. (c) Adaptive admittance controller with Demonstration-based task model.

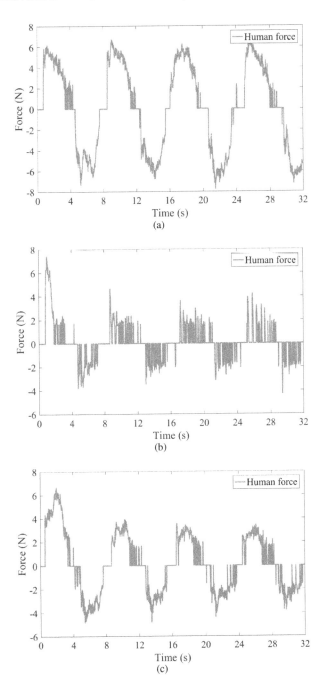

FIGURE 7.5 Human force: (a) Admittance controller with fixed parameters. (b) Adaptive admittance controller with linear task model. (c) Adaptive admittance controller with demonstration-based task model.

From the results, we can find that the admittance controller with fixed parameters requires a relatively large human force in the interaction, while the adaptive admittance controller requires a relatively small force. We can also find that the output error of the admittance model with demonstration-based task model is smaller than that of the admittance model with linear task model. And the continuity of the human force is better, which can provide a better user experience for the human tutor.

7.6.2 TEST OF THE NN-BASED CONTROLLER

In this group of experiments, the performance of NN learning is tested, which compensates for the uncertain dynamics of manipulator caused by the payload. The experiments are performed on the Baxter robot, which has two seven-degrees-of-freedom arms. The joints from the shoulder to end-effector are named $s0$, $s1$, $e0$, $e1$, $w0$, $w1$, and $w2$ in this chapter. The payload is attached using the left gripper of the robot, which weighs 0.94 kg. The robot is required to track a circular trajectory defined as $[X, Y, Z] = [0.65 + 0.1\sin(2\pi t/4), 0.2 + 0.1\cos(2\pi t/4), 0.2]$(m) with the orientation fixed. The corresponding trajectories in joint space, which are taken as the inputs of the proposed controller, are obtained through the inverse kinematics

We select three nodes for each input dimension of NN, and the centers of the nodes are distributed evenly within the limits of the joint position and velocity. There are $N = 2187$ NN nodes selected for $\hat{M}(\theta)$ and $\hat{G}(\theta)$, $2N$ nodes for $\hat{C}(\theta, \dot{\theta})$, and $4N$ nodes for $\hat{r}(\theta, \dot{\theta}, v_d, \dot{v}_d)$. In addition, the NN weight matrices are initialized as $\hat{W}_M = \mathbf{0} \in R^{nN \times n}$, $\hat{W}_C = \mathbf{0} \in R^{2nN \times n}$, $\hat{W}_G = \mathbf{0} \in R^{nN \times n}$, and $\hat{W}_r = \mathbf{0} \in R^{4nN \times n}$ with $n = 7$. The parameters of the error transformation function are set as $\delta_0 = 0.2$, $\delta_\infty = 0.04$, $a = 1$, and $\sigma = 1$.

The manipulator is controlled by the controller without NN learning in the first experiment, and the actual joint angles are recorded. Then, the controller with the proposed NN learning is employed to control the manipulator in the second experiment. The reference and actual joint angles in both experiments are shown in Fig. 7.6, and the corresponding tracking errors are shown in Fig. 7.7. The tracking errors are relatively high when NN learning is disabled, which is caused by the payload that the gripper holds, while in the second experiment, each joint of the manipulator tracks the reference trajectory very well and all tracking errors reduce into the interval $[-0.04, 0.04]$(rad) with the compensation torque increasing, which is shown in Fig. 7.8 (a). The gravity term of the manipulator dynamics is the main part that the payload affects, and hence, we particularly show the norm of each column of $\hat{W}_G$ in Fig. 7.8 (b). We can see that the norm of each column vector of the weight matrix $\hat{W}_G$ rises incrementally because of the torque generated by NN, and it still cannot compensate for the effect of the unknown dynamics. However, the rising speeds of all norms decrease with the increment in torque compensation.

7.6.3 POURING TASK

The second group of experiments aims to validate the DMP-based motion model. The ability of generalization and learning performance is tested. In these experiments, the demonstrations are performed by guiding the robot manipulator.

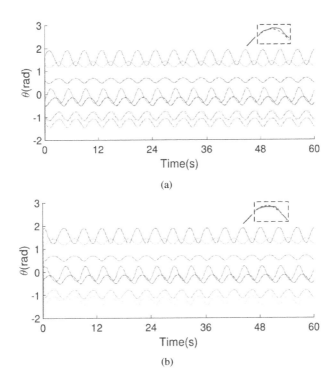

FIGURE 7.6 Reference joint angles (dashed lines) and actual joint angles (solid lines) when NN learning is (a) disabled, (b) enabled. The lines of different colors represent different joints.

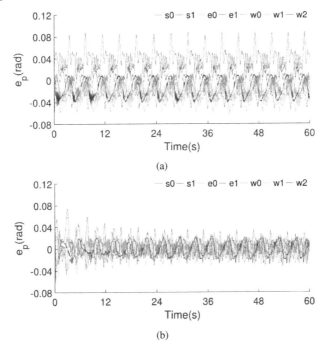

FIGURE 7.7 Tracking errors when NN learning is (a) disabled, (b) enabled.

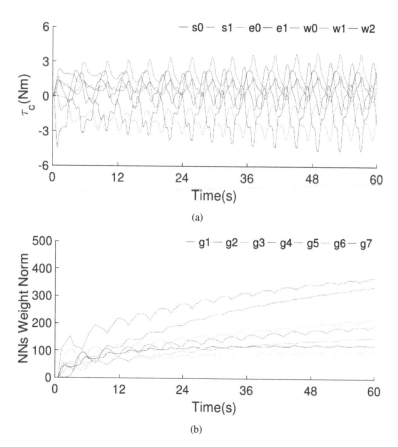

FIGURE 7.8 (a) The compensation torque. (b) The norm of each column of the weight matrix $\hat{W}_G$.

1) Ability of generalization: In this experiment, the tutor demonstrates how to pour water into a cup placed on the table, as shown in Fig. 7.9. The demonstration process is repeated five times. The joints $w0$, $w1$, $s0$, and $e1$ are moved while the others are fixed. The joint angles are recorded and used for learning the modified DMP. The parameters of the DMP model are set as $\tau_s = 1$, $l_1 = 25$, $l_2 = 10$, $\alpha_1 = \alpha_2 = 8$.

The learning results are shown in Fig. 7.10. The motions of the four joints are reproduced from the demonstrations, which synthesize the features of these demonstrations and enable the robot to complete the pouring task successfully, as shown in Fig. 7.12 (a). Subsequently, the target of the motion is modulated to the other cup. As shown in Fig. 7.11, the movement trajectory of each joint angle converges to the new goal, and the profile of each reproduction is retained. We test the generalized motion on the robot, and as shown in Fig. 7.12 (b), the robot can pour water into the other cup.

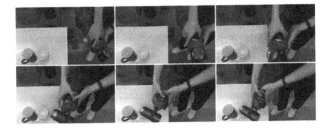

FIGURE 7.9 The demonstration process of a pouring task.

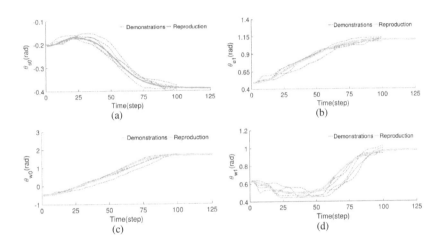

FIGURE 7.10 The learning results using the DMP-based motion model in a pouring task of (a) joint $s0$, (b) joint $e1$, (c) joint $w0$, and (d) joint $w1$.

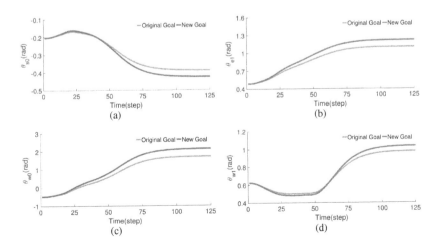

FIGURE 7.11 The generalization results using the DMP-based motion model in a pouring task of (a) joint $s0$, (b) joint $e1$, (c) joint $w0$, and (d) joint $w1$.

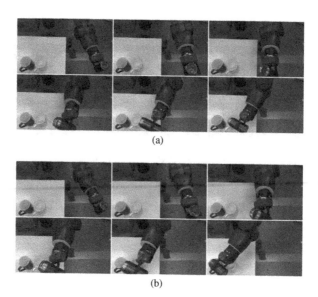

FIGURE 7.12 (a) The robot performs the pouring task with the regenerated motion. (b) The robot pours water into the other cup with the generalized motion.

7.7 CONCLUSION

This chapter has proposed a novel cognitive learning framework for human-robot skills transfer, which simultaneously considers the motion and the contact force during the demonstration. The DMPs are built to model the motion and the force to achieve skills generalization. The NN-based controller is designed to overcome the impact of the unknown payload so that the manipulator is able to track the given motions more accurately. The adaptive admittance model is proposed to simplify the teaching process. To reproduce the motion and the contact force, the hybrid force/motion controller is developed based on the original position controller of the Baxter robot. The NN-based controller is designed to overcome the impact of the unknown payload so that the manipulator is able to track the given motions more accurately. Experiments with the Baxter Robot have verified that the robot can perform the force-related task better than the motion-only method by employing the proposed robot learning framework. The success rate of task performance is also improved.

REFERENCES

1. Isura Ranatunga, Sven Cremer, Dan O Popa, and Frank L Lewis. Intent aware adaptive admittance control for physical human-robot interaction. In *2015 IEEE International Conference on Robotics and Automation (ICRA)*, pages 5635–5640. IEEE, 2015.
2. S Mohammad Khansari-Zadeh and Aude Billard. Learning stable nonlinear dynamical systems with Gaussian mixture models. *IEEE Trans. Robot.*, 27(5):943–957, 2011.

3. Heiko Hoffmann, Peter Pastor, Dae-Hyung Park, and Stefan Schaal. Biologically-inspired dynamical systems for movement generation: Automatic real-time goal adaptation and obstacle avoidance. In *2009 IEEE International Conference on Robotics and Automation*, pages 2587–2592, May 2009.

4. Adam Coates, Pieter Abbeel, and Andrew Y Ng. Learning for control from multiple demonstrations. In *Proc. IEEE Int. Conf. Mach. Learning*, pages 144–151, 2008.

5. Sylvain Calinon, Florent Guenter, and Aude Billard. On learning, representing, and generalizing a task in a humanoid robot. *IEEE Trans. Syst., Man, Cybern. B, Cybern.*, 37(2):286–298, 2007.

6. Zhaojie Ju and Honghai Liu. Fuzzy Gaussian mixture models. *Pattern Recognition*, 45(3):1146–1158, 2012.

7. Baibo Zhang, Changshui Zhang, and Xing Yi. Active curve axis Gaussian mixture models. *Pattern recognition*, 38(12):2351–2362, 2005.

8. Toshio Tsuji and Yoshiyuki Tanaka. Tracking control properties of human-robotic systems based on impedance control. *IEEE Transactions on systems, man, and cybernetics-Part A: Systems and Humans*, 35(4):523–535, 2005.

9. Christian Ott, Ranjan Mukherjee, and Yoshihiko Nakamura. Unified impedance and admittance control. In *Robotics and Automation (ICRA), 2010 IEEE International Conference on*, pages 554–561. IEEE, 2010.

10. Brenna D Argall and Aude G Billard. A survey of tactile human–robot interactions. *Robotics and autonomous systems*, 58(10):1159–1176, 2010.

11. Fanny Ficuciello, Raffaella Carloni, Ludo C Visser, and Stefano Stramigioli. Port-hamiltonian modeling for soft-finger manipulation. In *Proc. IEEE/RSJ Int. Conf. Intell. Robots Syst.*, pages 4281–4286, 2010.

12. Long Cheng, Zeng-Guang Hou, Min Tan, and Wen-Jun Zhang. Tracking control of a closed-chain five-bar robot with two degrees of freedom by integration of an approximation-based approach and mechanical design. *IEEE Trans. Syst., Man, Cybern. B, Cybern.*, 42(5):1470–1479, 2012.

13. Fan Ke, Zhijun Li, Hanzhen Xiao, and Xuebo Zhang. Visual servoing of constrained mobile robots based on model predictive control. *IEEE Trans. Syst., Man, Cybern. A, Syst.*, 47(7):1428–1438, 2016.

14. Chih-Min Lin and Ya-Fu Peng. Adaptive CMAC-based supervisory control for uncertain nonlinear systems. *IEEE Trans. Syst., Man, Cybern. B, Cybern.*, 34(2):1248–1260, 2004.

15. Sesh Commuri, Sarangapani Jagannathan, and Frank L Lewis. CMAC neural network control of robot manipulators. *J. Robot. Syst.*, 14(6):465–482, 1997.

16. Chenguang Yang, Yiming Jiang, Zhijun Li, Wei He, and Chun-Yi Su. Neural control of bimanual robots with guaranteed global stability and motion precision. *IEEE Transactions on Industrial Informatics*, 13(3):1162–1171, 2016.

17. Sepehr Saadatmand, Sima Azizi, Mohammadamir Kavousi, and Donald Wunsch. Autonomous control of a line follower robot using a q-learning controller. In *2020 10th Annual Computing and Communication Workshop and Conference (CCWC)*, pages 0556–0561. IEEE, 2020.

18. Aliasgar Morbi, Mojtaba Ahmadi, Adrian DC Chan, and Robert Langlois. Stability-guaranteed assist-as-needed controller for powered orthoses. *IEEE Transactions on Control Systems Technology*, 22(2):745–752, 2013.

19. Hsien-I Lin. Design of an intelligent robotic precise assembly system for rapid teaching and admittance control. *Robotics and Computer-Integrated Manufacturing*, 64:101946, 2020.

20. BR Manju and MR Sneha. ECG denoising using Wiener filter and Kalman filter. *Procedia Computer Science*, 171:273–281, 2020.
21. Isura Ranatunga, Frank L Lewis, Dan O Popa, and Shaikh M Tousif. Adaptive admittance control for human–robot interaction using model reference design and adaptive inverse filtering. *IEEE transactions on control systems technology*, 25(1):278–285, 2016.
22. Peter Pastor, Heiko Hoffmann, Tamim Asfour, and Stefan Schaal. Learning and generalization of motor skills by learning from demonstration. In *2009 IEEE International Conference on Robotics and Automation*, pages 763–768. IEEE, 2009.
23. Thomas Cederborg, Ming Li, Adrien Baranes, and Pierre-Yves Oudeyer. Incremental local online Gaussian mixture regression for imitation learning of multiple tasks. In *2010 IEEE/RSJ International Conference on Intelligent Robots and Systems*, pages 267–274. IEEE, 2010.
24. Hsi Guang Sung. *Gaussian mixture regression and classification*. PhD thesis, Rice University, 2004.
25. Bin Ren, Xurong Luo, Yao Wang, and Jiayu Chen. A gait trajectory control scheme through successive approximation based on radial basis function neural networks for the lower limb exoskeleton robot. *Journal of Computing and Information Science in Engineering*, 20(3), 2020.
26. Yafeng Li, Aimin An, Jinlei Wang, Haochen Zhang, and Fancheng Meng. Adaptive robust of RBF neural network control based on model local approximation method for upper limb rehabilitation robotic arm. In *Journal of Physics: Conference Series*, volume 1453, page 012086, 2020.
27. Chenguang Yang, Xingjian Wang, Zhijun Li, Yanan Li, and Chun-Yi Su. Teleoperation control based on combination of wave variable and neural networks. *IEEE Transactions on Systems, Man, and Cybernetics: Systems*, 47(8):2125–2136, 2016.
28. Zhaoyong Mao and Fuliang Zhao. Structure optimization of a vibration suppression device for underwater moored platforms using CFD and neural network. *Complexity*, 2017, 2017.
29. Zaiji Piao, Chen Guo, and Shuang Sun. Adaptive backstepping sliding mode dynamic positioning system for pod driven unmanned surface vessel based on cerebellar model articulation controller. *IEEE Access*, 8:48314–48324, 2020.
30. Zhijun Li, Jun Deng, Renquan Lu, Yong Xu, Jianjun Bai, and Chun-Yi Su. Trajectory-tracking control of mobile robot systems incorporating neural-dynamic optimized model predictive approach. *IEEE Transactions on Systems, Man, and Cybernetics: Systems*, 46(6):740–749, 2015.

Index

Italicized and **bold** pages refer to figures and tables respectively